Plastic Injection Molding

Plastic Injection Molding

...mold design and construction fundamentals

By Douglas M. Bryce

Volume III: *Fundamentals of Injection Molding* series

SME

Published by the
Society of Manufacturing Engineers
Dearborn, Michigan

FUNDAMENTALS OF INJECTION MOLDING SERIES

Volume I: *Manufacturing Process Fundamentals*
Library of Congress Card Number: 96-067394
International Standard Book Number: 0-87263-472-8

Volume II: *Material selection and Product Design Fundamentals*
Library of Congress Card Number: 97-068807
International Standard Book Number: 0-87263-488-4

Volume III: *Mold Design and Construction Fundamentals*
Library of Congress Card Number: 98-060567
International Standard Book Number: 0-87263-495-7

Additional copies may be obtained by contacting:

Society of Manufacturing Engineers
Customer Service
One SME Drive, P.O. Box 930
Dearborn, Michigan 48121
(800) 733-4763; fax (313) 271-2861
http://www.sme.org

SME staff who participated in producing this book:

Philip E. Mitchell, Handbook Editor
Rosemary K. Csizmadia, Production Team Leader
Karen M. Wilhelm, Manager, Book Publishing
David K. McWilliams, Production Assistant
Cover design by Judy D. Munro, Manager, Graphic Services

Printed in the United States of America

Table of Contents

Chapter 3 – Basics of Mold Construction

Chapter 4 – Action Areas of the Mold

Chapter 5 – Runners, Gates, and Venting

Chapter 6 – Controlling Mold Temperatures

Chapter 7 – Mold Alignment Concepts

Chapter 8 – Repairing, Protecting, and Storing Molds

Chapter 9 – Troubleshooting Product Defects Caused by Molds

List of Tables and Figures

Chapter 1

Chapter 2

Chapter 3

Chapter 4

Chapter 5

Chapter 6

Chapter 7

Chapter 8

Chapter 9

Preface

This book is the third in a series providing basic information, concepts, and ideas to those interested in, or already existing in, the world of injection molding of thermoplastics (Volume I covers manufacturing process fundamentals and Volume II explains material selection and product design fundamentals). This volume covers the subject of fundamental mold design and construction. My intention was to write this book (and the others in the series) in such a way that both newcomers and old-timers would be able to obtain information that is otherwise not readily available.

In this book we take a look at the role of the mold in the injection molding process, and how it should be designed and built. Along the way we discuss mold components, various materials from which to make molds, and some of the more popular mold designs. Of course, we also look at some of the methods and equipment used for making these molds. We'll also address a few of the important design criteria that are required for both product and mold design.

Whether you are involved in designing and/or making molds or not, this book is intended for you. Please enjoy reading it as much as I enjoyed writing it. I would love to hear from you if you can take the time to write and tell me about your experiences in the injection molding industry. In fact, maybe you should consider writing a book yourself. Good luck!

I would like to take a moment to thank the following for their contribution to the completion of this volume.

- *Society of Manufacturing Engineers*, for providing the various resources for editing and publishing the entire series.
- The *reviewers* of this book before publication, for giving valuable advice on how to make it better:

    Lawrence G. Cook, President, MidWest Plastics Sales & Associates
    Marion Cooper, President, Cooper Engineering, Inc.
    Jon H. Eickhoff, C.E.O. Ronningen Research and Development Company

- *Texas Plastic Technologies*, for supporting my efforts and continuing operations smoothly during those few times when I needed to be locked up in my office writing.
- *Various moldmakers and mold designers*, who shared design and construction ideas with me as well as their thoughts and opinions of what I was trying to do and say.

I am dedicating this volume to my father, from whom I inherited the desire, energy, and perseverance to complete such an achievement, and for being there to provide a gentle prod when needed.

You may contact me through Texas Plastic Technologies at the following address:

Texas Plastic Technologies
605 Ridgewood Road West
Georgetown, Texas 78628
Phone: (512) 863-5933
Fax: (512) 869-2680
E-mail: dbryce@texplas.com
Web site: http://www.texplas.com

Douglas M. Bryce
Georgetown, Texas 1998

A Primer on Molds

1

THE INJECTION MOLDING CONCEPT

In its simplest form, the injection molding process is like the operation of a hypodermic needle. A barrel contains heated plastic that is injected (by use of a plunger or auger device) into a closed mold that contains a machined, reverse image of the desired product, as shown in Figure 1-1. This image is called the *cavity image*.

The injected plastic is allowed to cool and solidify in the cavity. Then, the mold is opened and the product is ejected. While this may seem simple, the process actually involves many individual activities and parameters that must be tightly controlled to produce a high-quality product at a reasonable cost.

The primary advantage of this process is that many functions and features can be incorporated into the product design. This process will minimize, or eliminate, the amount of secondary work required to produce the same product in other ways or using other materials, as shown in Figures 1-2 and 1-3.

In addition, the process is fast, usually requiring less than one minute for a completed product. If multiple cavity images can be placed in the mold, many products can be produced at the same time. This makes the cost of an individual part much lower than if it was molded alone.

Now, let's take a look at the importance of the mold in the injection molding process.

IMPORTANCE OF THE MOLD

The plastic product begins life in the mold. The mold produces the final shape of the product before it is ejected out into the world to perform its intended function. In theory, the mold can be designed and built to create a totally finished product, including painted surfaces, assembled units from individual components, and molded-in metal inserts. However, if more features and functions are required from the mold, then costs and time will increase during its construction. Therefore, a great deal of thought should go into exactly what is required from the mold before the first mold design is even considered.

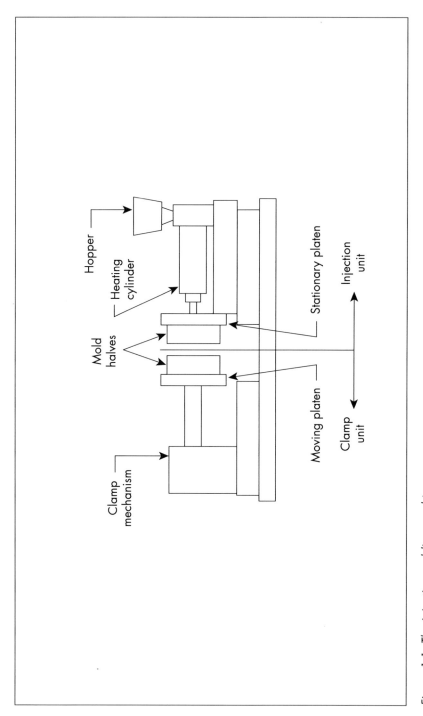

Figure 1-1. The injection molding machine.

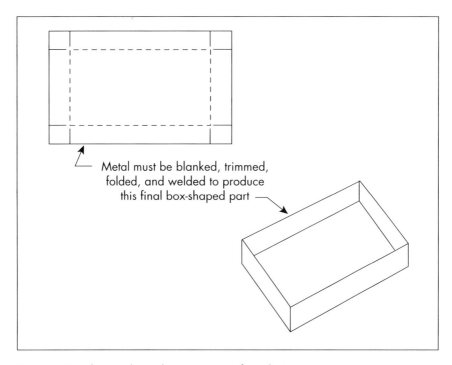

Metal must be blanked, trimmed,
folded, and welded to produce
this final box-shaped part —

Figure 1-2. Fabricated metal components of product.

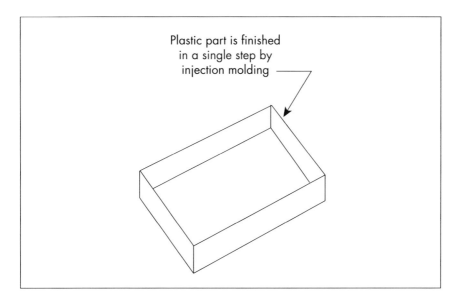

Plastic part is finished
in a single step by
injection molding —

Figure 1-3. Plastic molding with integrated components.

MOLD BASE COMPONENTS

A mold is constructed using a series of components including various plates, pins, bushings, pillars, ejector systems, and many other items used for many purposes. Figure 1-4 shows some of the basic items and where they are located in the mold.

Figure 1-4 shows the mold in the closed position. When the mold is open it separates between the A and B plates. This plane is called the *parting line* because it designates where the mold parts or separates. Figure 1-5 shows the mold in this separated stage.

We refer to each half of the separated mold by whether it contains the A plate or the B plate. Therefore, the half of the mold containing the A plate is called the A half, and the half containing the B plate is called the B half. We also use the term *live half* for the B half because it usually contains the moving section known as the *ejector system*. Normally, the A half does not contain any moving sections, so it is referred to as the *dead half*.

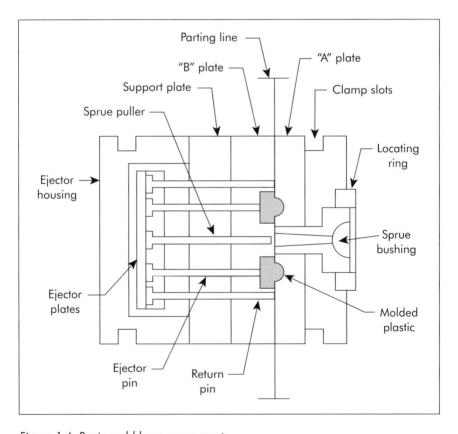

Figure 1-4. Basic mold base components.

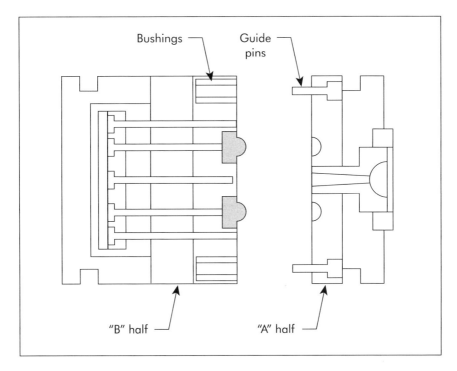

Figure 1-5. Mold open at the parting line.

The Injection Half of the Mold

This section will discuss the sprue bushing, A and B plates, runners and gates, and cavity image.

The Sprue Bushing

In most cases, the A half contains the sprue bushing. The term *sprue* was borrowed from the metal die-casting industry, and apparently comes from a Celtic (Scottish) word meaning *to spread*. The sprue bushing in an injection mold is the component that allows molten plastic to enter the mold and begin its travel to the cavity image. Figure 1-6 shows a cutaway of how this occurs.

Figure 1-6 shows the molten plastic leaving the injection press and entering the mold through the sprue bushing. The sprue bushing directs the material through the runner system, past the gate, and into the cavity image. The sprue bushing is the interface between the injection mold and the injection machine. It is made from hardened tool steel, highly polished to minimize sticking, and is a replaceable component of the mold.

Typically, the critical dimensions of a sprue bushing are determined by the viscosity of the molding material, the volume of material traveling through the

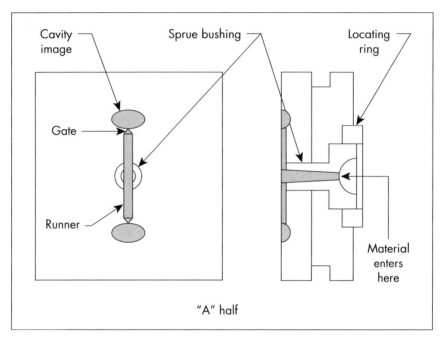

Figure 1-6. Function of sprue bushing.

runner system, and the thickness of the mold plates of the A half of the mold. In addition, the diameter of the orifice of the nozzle is a factor. Figure 1-7 shows the concept behind the dimensioning of a typical sprue bushing.

The A and B Plates

The mold components that form the molded part are the A and B plates. To take advantage of the ejector system (usually located in the B half of the mold) we mold the plastic product in the B plate whenever possible. However, there are some situations in which the plastic product design requires some of the cavity image to be placed in the A plate, such as in a three-plate mold which we will discuss later. Figure 1-8 shows the A and B plates from a standard two-plate mold used to make triangular plastic safety signs. The mold is a two-cavity mold, meaning two pieces are molded at the same time.

Runners and Gates

After the molten plastic leaves the injection molding machine and enters the mold through the sprue bushing, it must be directed to the cavity image and allowed to enter the cavity image to form the finished part. This is accomplished by using a runner system and gates, as shown in Figure 1-9. The runner system is a pathway

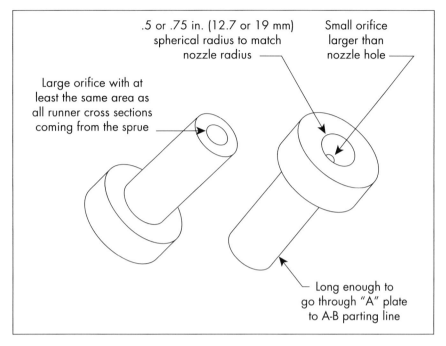

Large orifice with at
least the same area as
all runner cross sections
coming from the sprue

.5 or .75 in. (12.7 or 19 mm)
spherical radius to match
nozzle radius

Small orifice
larger than
nozzle hole

Long enough to
go through "A" plate
to A-B parting line

Figure 1-7. Dimensioning a sprue bushing.

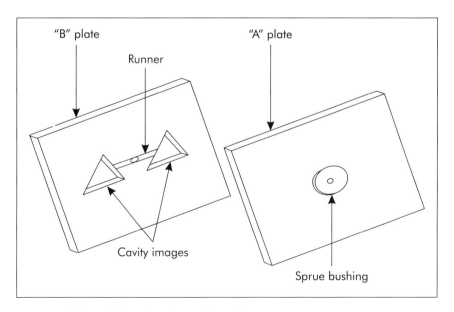

"B" plate

Runner

"A" plate

Cavity images

Sprue bushing

Figure 1-8. Typical A and B plates of a mold.

machined into the face of the mold base. Usually this is accomplished by use of a ball end mill, but it also can be created using electrical discharge machining (EDM).

The gate is located at the end of the runner system and is considered an opening in the mold, like a window. It is designed and constructed to allow molten plastic to enter the cavity at proper velocity and volume needed to fill the cavity quickly, but in a controlled condition. The thickness of the gate has a major function in determining the overall cycle of the injection molding process and must be kept as thin as possible while allowing proper flow.

While it is theoretically possible to fill any size or shape cavity image with only a single gate, sometimes it is necessary to utilize the flow profile and characteristics of multiple gate systems. These will be discussed later.

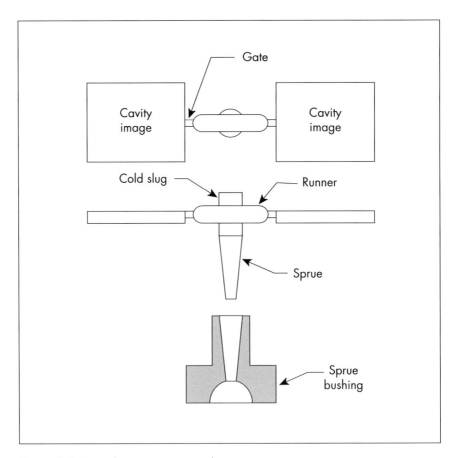

Figure 1-9. Typical runner system and gates.

The Cavity Image

The shape of the final molded product is determined by the shape of the cavity image that is machined into the A and B plates (as well as the characteristics of the specific material being molded). This image is machined into the plates using common machining methods such as milling, grinding, lathe-working (turning), and stoning/polishing. But, the most common method of cavity image machining today is EDM (also known as EDNC or electrical discharge numerical control). This method utilizes electrical current to erode metal from the cavity plates. It can be controlled to the point of requiring no further machining if the cavity image does not need a high-polish finish. EDM methods are discussed in another chapter.

The Ejector Half of the Mold

This section will describe the ejector housing, support plate, support pillars, ejector pins, return pins, sprue pullers, and ejector system guide pins and bushings.

The Ejector Housing

The ejector housing is a U-shaped metal box, formed as a forging in a solid, one-piece unit. This is done to attain maximum strength and minimum deflection during molding, as shown in Figure 1-10.

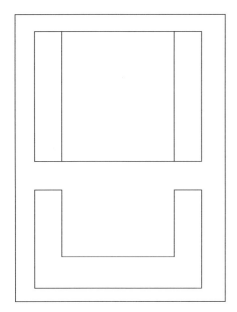

To cut corners and save money, some moldmakers attempt to fabricate the housing from bolted plates. This is an unsafe practice because the side rails of the box will distort under injection and clamp pressure and the box may crack. In addition, the distortion causes excessive wear on the ejector pins and the holes in which the pins ride. It is questionable whether or not any money is actually saved by this fabrication process, certainly when the potential danger and damage is considered.

The *ejector housing* (commonly called the *ejector box*) is used to contain the ejector system. This system consists of ejector pins, return pins, sprue-puller pins, and possibly many springs, bushings, and guide pins. Guided ejector systems utilize guide pins and bushings to minimize wear

Figure 1-10. The U-shaped ejector housing.

on the ejector system components. Although this practice adds slightly to the overall cost of the mold, it does minimize repair costs that occur due to ejector system wear.

Support Plate

Because the ejector housing is U-shaped, there is a great amount of unsupported area directly under the B plate of the mold. This open area allows the B plate to distort by bowing or bending under injection pressure. The amount of distortion is about 0.010 in. (0.254 mm) or more. This is enough to cause a lot of flash on the parting line and will quickly result in a weakened B plate. To keep this from happening, a support plate is mounted directly behind the B plate, as shown in Figure 1-11. The use of two plates (the B plate and the support plate) creates a much greater resistance to distortion than increasing the thickness of the B plate.

Support Pillars

In addition to the support plate, support pillars are also used to help reduce (or eliminate) the B plate distortion, as shown in Figure 1-12. These are round steel

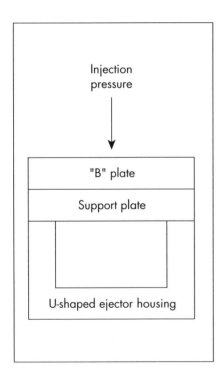

Figure 1-11. Support plate.

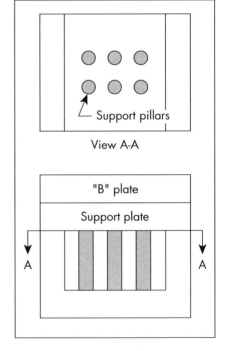

Figure 1-12. Support pillars.

columns that are placed as props between the ejector housing and the support plate of the mold.

Ejector Pins, Return Pins, and Sprue Pullers

Ejector pins are steel "nails" with a head, body, and flat face. These pins are used to push the finished, molded product out of the mold at the end of the injection molding cycle. Ejector pins are also used to eject the surface-style runner system. These pins are called ejector pins only if they contact the plastic surface of the molded product or runner, as shown in Figure 1-13.

Where there is an ejector pin, there is going to be an impression showing on the molded part. This impression will take the form of either a depression in the plastic or a raised pad of excess material. This is discussed in more detail in another chapter.

Return pins are used for pushing the ejector system back into its proper place when the mold closes in preparation for the next molding cycle. These pins differ from ejector pins in that they do not contact the surface of the molded part, but rather contact the steel surface of the A plate as the mold closes. This contact forces the return pins to push back the ejector system, thus returning it to the proper location.

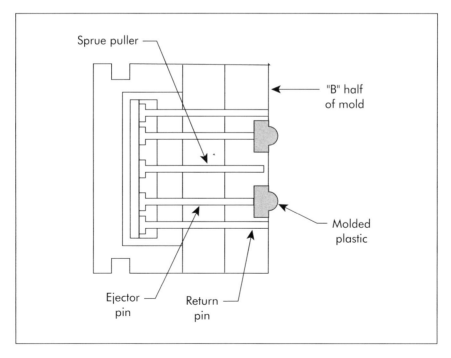

Figure 1-13. Ejector pins, return pins, and sprue pullers.

Sprue puller pins are more elaborate ejector pins. Their primary purpose is to assist in releasing the sprue from the sprue bushing when the mold first opens after the injection process is complete. The sprue puller pin usually incorporates an undercut of some type on its end. This undercut can take the shape of a groove machined into the pin diameter, a Z-shaped notch cut into the diameter, or even a ring groove cut into the hole in which the pin rides. In all cases, the undercut traps material from the sprue. When material in the undercut hardens, it causes the sprue to be pulled from the sprue bushing as the mold opens up. After the mold opens all the way, the ejector system activates and the sprue puller acts as an ejector pin to push the sprue undercut off the B half of the mold.

Ejector System Guide Pins and Bushings

Because the mold is usually operating in a horizontal plane, gravity tends to pull the ejector system downward. This action will cause undue wear on the diameters of all the pins as well as the holes in which they ride. This will cause an elliptical hole to form in the mold and flash will fill it. One way to minimize the effects of gravity is to use guided ejector systems, as shown in Figure 1-14. These systems utilize guide pins and bushings that are designed to overcome the gravity effects and keep the ejector system operating on a true horizontal plane at all times.

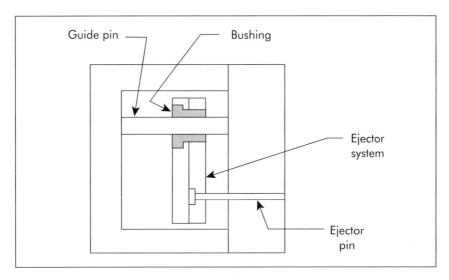

Figure 1-14. Ejector system guide pins and bushings.

SUMMARY

In its simplest form, the injection molding process is like the operation of a hypodermic needle. A barrel contains heated plastic that is injected (by use of a plunger or auger device) into a closed mold that contains a machined, reverse image of the desired product. The primary advantage of this process is that many functions and features can be incorporated into the product design. This process will minimize, or eliminate, the amount of secondary work required to produce the same product in other ways or using other materials.

The plastic product begins life in the mold. A mold is constructed using a series of components, including various plates, pins, bushings, pillars, ejector systems, and a multitude of other items used for many purposes.

The mold components that form the molded part are the A and B plates. The shape of the final molded product is determined by the shape of the cavity image that is machined into the A and B plates (as well as the characteristics of the specific material being molded).

The ejector housing is a U-shaped metal box, formed as a forging in a solid, one-piece unit. The ejector housing (commonly called the ejector box) is used to contain the ejector system. This system consists of ejector pins, return pins, sprue-puller pins, and possibly many springs, bushings, and guide pins.

QUESTIONS

1. What is the name of the mold area that contains a machined, reverse image of the desired plastic product?
2. What is the primary advantage of injection molding?
3. The injection molding process is fast, usually requiring _____ for a completed product.
4. Where does a standard mold separate when it opens?
5. What is the purpose of the sprue bushing?
6. How is the molten plastic directed from the sprue bushing to the cavity images?
7. Theoretically, how many gates are needed to fill any size or shape cavity image?
8. Why is it an unsafe practice to attempt to fabricate an ejector housing from plates bolted together?
9. What is the primary difference between an ejector pin and a return pin?

Mold Design Basics

2

GATHERING INFORMATION

Before a mold design is started, there are some basic facts that need to be gathered. These include determining how many cavities to build, what material to make the mold out of, and other data that we will discuss in this section. It is probable that more than one area of expertise may be required to obtain all the information that is necessary. This may result in soliciting specialized assistance from others such as material engineers and financial analysts. However, we can make some basic assumptions, based on whatever knowledge and information is available at the time.

Which Plastic?

While the mold designer usually does not select the molding material, he or she should be aware of the more important aspects and characteristics involved in molding specific plastics. For example, shrinkage factors sometimes vary widely between different materials and may vary among different grades and versions of the same material. Also, some plastics will absorb and dissipate heat more efficiently than others, resulting in more efficient cooling during the molding process. This may affect cooling channel locations in the mold, and the viscosity of a particular plastic has a large influence on gate, runner, and vent design, location, and construction.

A thorough study of the characteristics of various plastics is not possible within the scope of this book, and the reader is advised to locate such information in Volume 2 of this series: *Material Selection and Product Design Fundamentals*. Instead, we will discuss the basic information that is desirable to know about a specific material as it applies to mold design and construction.

Determining Shrinkage

Every material we know of (except water) expands when it is heated and contracts when it is cooled. In the field of plastics, we define the contraction phase as *shrinkage*. Each plastic material has a *shrinkage factor* (sometimes incorrectly called a shrink rate) assigned to it. This factor is used to estimate how much a part will shrink after it is removed from the mold. After that is determined, the mold can be built to a set of dimensions that create a molded part large enough so that it will contract to the desired finished size after shrinkage.

Those plastics that shrink equally in all directions (notably amorphous materials) are referred to as having *isotropic shrinkage*. Some plastics (notably crystalline materials) will shrink more in the direction of flow than across the direction of flow (unless they are reinforced, in which case, shrinkage is greater across the direction of flow). This type of shrinkage, which is not equal in all directions, is known as *anisotropic shrinkage*. Both types of shrinkage are shown in Figure 2-1.

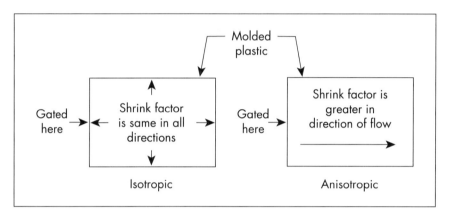

Figure 2-1. Part shrinkage.

Shrink factors are assigned on an inch-per-inch basis, meaning that the factor is applied to every inch (or fraction of inch) of every dimension of the product to be molded. For example, if a product is to be 6 in. (15.2 cm) in length, and the shrink factor is .010 in./in. (0.010 cm/cm), the mold cavity must be made to be 6.060 in. (15.4 cm) in length to produce the 6-in. (15-cm) long product.

The shrink factors are rated as low, medium, or high. *Low shrink* is commonly accepted as from .000 in./in. to .005 in./in. (0.000 cm/cm to 0.005 cm/cm). *Medium shrink* is commonly accepted as from .006 in./in. to .010 in./in. (0.006 cm/cm to 0.010 cm/cm), and *high shrink* is commonly accepted as anything over .010 in./in. (0.010 cm/cm). Some plastics may have as great as .075 in./in. (0.075 cm/cm) shrinkage. Amorphous materials tend to exhibit low shrink, semicrystalline materials have medium shrink, and crystalline materials tend to show high shrinkage. If glass reinforcement is added to the plastic, the shrinkage will be less than that same plastic in a *neat* (no reinforcement added) condition. That's because the nonshrinking glass reinforcement takes up some of the volume of the mass of plastic and dilutes the shrinkage of the total mass.

Shrinkage is difficult to estimate because there are many items that influence the final shrinkage result. Changes in wall thickness of the product design may cause different shrinkage to occur in certain areas of the molded part. Temperature variations in the mold (greater than $10°$ F [$5.5°$ C] between any two points) may result in varying shrinkage areas of the molded part. The mold designer can only use a best guess method of determining shrinkage by gathering as much advice from experienced molders, moldmakers, material suppliers, and other designers as possible, and then staying steel-safe on all dimensions to allow adjustments after sampling the mold.

It is not unusual to have to *develop* the mold according to the shrinkage idiosyncrasies of the molding material. For example, in a case where a hole is molded by using a core pin fastened in the mold, the core pin will be made to the part dimension diameter plus shrinkage. But, the finished product may be molded with the hole having an elongated diameter rather than a perfectly round one, due to shrinkage conditions. After the molder has successfully optimized the molding process, the core pin may have to be machined to be in an out-of-round condition so the final molded part, after cooling and shrinking, will actually have a truly round hole diameter. This development process might occur in other areas of the cavity because of the molding material's shrinkage characteristics and the product design.

Pressure and Viscosity

Viscosity is a measurement of the thickness of a material in its liquid (molten) state. The higher the viscosity, the thicker the material. A high viscosity plastic material requires a greater amount of injection pressure to push it through the mold than a low viscosity plastic. In addition, the high viscosity materials require larger runner diameters and greater gate volume to allow easy flow to the cavity image. And, the high viscosity plastics allow deeper vents for faster removal of trapped air.

The viscosity of a plastic determines how much pressure will be needed to inject the material into a mold. Viscosity is measured by elaborate and relatively expensive test equipment. But, it can be indicated inexpensively by using an ASTM test D1238, which uses a small amount of plastic material and simulates the injection molding process, as shown in Figure 2-2. This test is called the *melt index* test, but is also called *melt flow*, *flow index*, and *melt rate*. A machine called a *plastometer* is programmed to a set of conditions dictated by the plastic being analyzed. Currently, there are at least 33 sets of conditions and each plastic will fall within one of these conditions.

The melt index number can be used as a tool for determining the flowability of a particular plastic. The test begins by dropping an amount of raw plastic in the heating chamber, placing a plunger device in the chamber, setting a predeter-

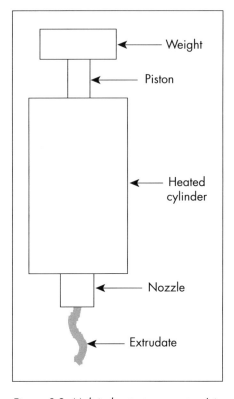

Figure 2-2. Melt index test apparatus (viscosity indicator).

mined load on top of the plunger, and measuring the amount of material that extrudes from the nozzle in a 10-minute period. The resultant number is the melt index value measured as grams per 10 minutes. Flow numbers (MI numbers) usually fall within a range of 2 to 50 with 12 to 14 being the most common (roughly half an ounce per 10 minutes). This means that 12 to 14 grams of plastic came from the orifice of the test machine in a total of 10 minutes.

The lower the melt index number, the stiffer the flow of the material. That means a higher injection pressure will be required to fill the mold, larger runner diameters will be needed for surface runner molds, and gate depths will probably increase. Vent depths are affected by viscosity; as the MI number gets smaller, the vents are made deeper to allow trapped air to escape the mold faster and reduce injection pressure requirements.

Viscosity also affects physical properties of the molded part. Basically, the higher the MI number within the range associated with a given plastic, the weaker the molded part. Conversely, the lower the MI, the stronger the part. Some of the ways that molded part properties are affected by viscosity are shown in Table II-1.

How Many Cavities?

Before we can determine the size of mold and the size of equipment needed to run the mold, we must determine how many cavities are required. Along with the total time of a cycle, the number of cavities determines how many molded parts can be produced during one complete cycle of the injection molding process. The number of cavities needed depends on the time frame established for producing the annual volume requirements of a specific product. For example, if an average of 100,000 units a year is required, we need to determine how many cavities are required to produce the product during the year. First, determine the production time available for the year. Most molding operations produce parts 24 hours a

Table II-1. Effect of Viscosity on Physical Properties
(Relationship as Melt Index Value Decreases)

Stiffness	Increases
Tensile strength	Increases
Yield strength	Increases
Hardness	Increases
Creep resistance	Increases
Toughness	Increases
Softening temperature	Increases
Stress-crack resistance	Increases
Chemical resistance	Increases
Molecular weight	Increases
Permeability	Decreases
Gloss	Decreases

Note: Permeability and gloss actually decrease *as the melt index value drops.*

day, five days a week. Weekends are used for maintenance. Assuming a 52-week year, 5 days a week, and 24 hours a day, we arrive at a total time of 6,240 hours a year. Each month then has an average of 520 hours available (6,240/12).

To calculate how many cavities we will need to machine into the mold, we will have to estimate a cycle time. The cycle time is determined primarily by the thickest wall section of the part. For a guideline, Figure 2-3 can be used to make this determination and assumes that the mold will be placed in a properly sized molding machine and that all phases of the injection process are average times. Different materials may result in longer or shorter times, but Figure 2-3 is the result of actual tests performed by Texas Plastic Technologies from 1991 through 1994.

Note that the chart line in Figure 2-3 does not rise at a straight angle. That is due to the cooling time portion of the overall cycle. As the wall thickness doubles, the cooling time actually increases four-fold. That is why it is beneficial to keep wall thicknesses (and gate thickness) at an absolute minimum.

After the total cycle time is estimated, using Figure 2-3, the number of cycles per hour can be calculated by dividing 3,600 (the number of seconds in an hour) by the estimated cycle time. Let's make an assumption that the part in question has a maximum wall thickness of .100 in. (25.4 mm). From Figure 2-3, we find that the total cycle time would be approximately 36 seconds. Dividing that number into 3,600 shows that we can mold 100 cycles per hour. Now, we can figure out how many cavities we need. If we have a single cavity mold we can produce 100 units per hour. That means it would take approximately 1,000 hours, or 8.33 weeks, to mold our annual requirement of 100,000 units. If we built a two-cavity mold we

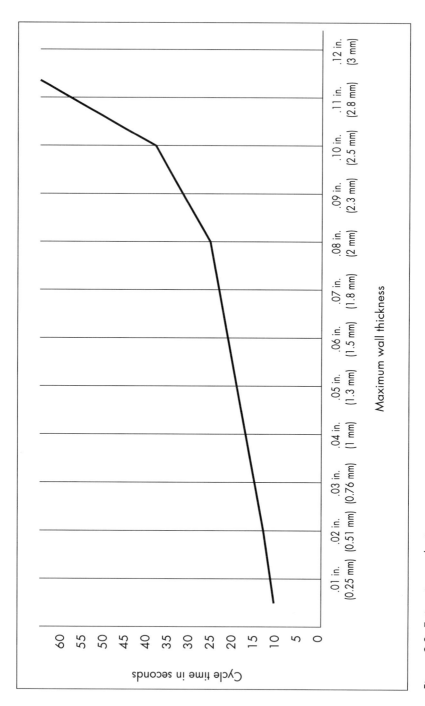

Figure 2-3. Estimating cycle times.

could mold the total 100,000 units in half the time, or 4.16 weeks. Of course, the two-cavity mold would be more expensive and this must be considered.

Now, let's consider a situation where we need 10 *million* units a year. If the cycle time stays at 36 seconds, we still have a total of 100 cycles per hour. Therefore, a single-cavity mold could make as many as 624,000 units if it ran all year long. We would need at least a 16-cavity mold to produce all 10 million parts on one mold, and that would mean running the mold constantly, even a few overtime hours on weekends. This is not normally practical. What we might consider is two to four molds, each with 16 to 32 cavities each. Then, we could produce the entire year's worth of requirements in 3 to 6 months molding time, depending on which combination of cavity and mold numbers we decided to use. Also, we would be utilizing two to four molding machines at once and that must be taken into consideration. If we do not have enough molding equipment on hand (or the proper size) we may have to farm out the molding, or we could purchase additional equipment if it can be justified. These are some of the issues that must be addressed before deciding how many cavities should be placed in a mold.

Which Mold Material?

There are dozens of materials that can be used for making molds for producing plastic products, including many types of aluminum, brass and copper, epoxy, and many others, as well as combinations of these. The following section describes some of the more common materials and the role they play in the making of molds.

Steels

1020 carbon steel. This steel is used for ejector plates and ejector retainer plates and is easily machined and welded. Not usually hardened because of distortion and warp, this material must be first carburized if hardening is preferred.

1030 carbon steel. Used for mold bases, ejector housings, and clamp plates, this steel has 25% greater tensile strength than 1020 and can be easily machined and welded. It can be hardened to Rockwell hardness C scale (R_c) 20 to 30.

1040 carbon steel. Commonly used for support pillars, this tough steel has good compressive strength and can be hardened to R_c 20 to 25.

4130 alloy steel. This is a high-strength steel used primarily for cavity and core retainer plates, support plates, and clamping plates, and is supplied at 26 to 35 R_c.

6145 alloy steel. Primary use for this type of steel is for sprue bushings and it is supplied at 42 to 48 R_c.

S-7 tool steel. Shock resistant with good wear resistance, this steel is used for interlocks and latches and hardened to 55 to 58 R_c.

O-1 tool steel. This is a general purpose, oil-hardening steel used for small inserts and cores and hardened to 56 to 62 R_c.

A-2 alloy tool steel. This steel has good dimensional stability and abrasion resistance, and is used for hobs and slides and is hardened to 55 to 58 R$_c$.

A-6 tool steel. A general purpose oil-hardening steel with good dimensional stability and high hardness, its primary use is for optical quality cavities and cores and it is hardened to 56 to 60 R$_c$.

D-2 tool steel. This steel has high chromium and high carbon content, and is difficult to grind, but has excellent abrasion resistance. It is used for gate inserts, lifters, and slides, and is hardened to 58 to 60 R$_c$.

H-13 tool steel. This is a very high toughness, low-hardness steel used for high quality cavity and core requirements. It is primarily used for ejector pins, return pins, sprue pullers, leader pins, and slide-actuating angle pins, and supplied annealed at 15 to 20 R$_c$, but can be hardened to 60 R$_c$ with little distortion.

P-20 tool steel. This is a modified 4130, commonly referred to as *prehard*. It is supplied at a R$_c$ hardness of 28 to 40, which provides moderately high hardness, good machinability, and exceptional polishability. It is used primarily for cavities and cores, as well as stripper plates.

420 stainless steel. Used in applications requiring exceptional chemical resistance (such as molding PVC resins), this steel is usually supplied in an annealed condition (15 to 25 R$_c$), but can be hardened to 55 to 60 R$_c$. Its primary use is as a steel for cores and cavities.

Aluminum

While there are many grades of aluminum available for making molds, the most common, and most efficient to work with, is the aircraft grade 7075 (T6). This wrought aluminum alloy is produced by hot rolling cast aluminum to the desired thickness of plate. The entire mold can be made of the same material (including cavity and core) and an anodizing process can be utilized to impart a surface hardness of up to 65 R$_c$ for wear resistance. However, due to the smoothing tendency of the normal aluminum surface, it is possible to mold with no surface treatment. The microscopic hills and valleys of the aluminum surface tend to even themselves out without galling. Use of 7075 aluminum can result in mold build times being reduced by up to 50% (due to faster machining times) and the injection molding cycle being reduced by up to 40% (due to faster heat dissipation), depending on size and complexity of the product being molded.

Until recently, aluminum was considered as a moldmaking material only for low-volume production or prototype molds. The use of 7075 alloy has created opportunities to use aluminum for high-volume production in up to millions of cycles. Even glass-reinforced and high-temperature plastics can be molded successfully in aluminum molds.

Beryllium-copper Alloys

High strength and high levels of thermal conductivity make the beryllium-copper alloys excellent selections for making cores and cavities for injection molds.

They are commonly used as components that are fitted to steel mold bases, but also can be used in conjunction with aluminum mold bases for greater economy. They are particularly useful for situations where the placement of cooling channels in the mold makes heat removal difficult, such as in deep draw parts or parts with unusual contours. Strategically placed beryllium-copper components will assist in dissipating the heat from these areas without using complicated water line channels.

The types of beryllium-copper most commonly used for cores and cavities are CuBe 10, CuBe 20, and CuBe 275. They differ mainly in tensile strength with the higher numbers having the greater strength. In addition, the higher number grades allow higher levels of hardness. This ranges from a low of Rb 40 for CuBe 10 to a maximum of R_c 46 for CuBe 275.

Other Materials

There are other materials that can be used for making molds for plastic injection molding, including epoxy, aluminum/epoxy alloys, silicone rubbers, and even wood. However, these are usually all reserved for very small volumes, such as under 100 pieces. In most cases, these represent prototype volumes, and the molds are not expected to meet the demanding requirements of higher-volume production levels. The scope of this type of mold construction does not fit the intent of this publication, and the reader requiring further information is encouraged to contact mold makers specializing in the prototype field. A listing of these can be obtained from the Society of the Plastics Industry (SPI), 1275 K Street N.W. Suite 400, Washington, DC 20005.

MOLD MATERIAL SURFACE ENHANCEMENTS

Surface enhancement methods and processes are utilized to impart specific properties to the mold, usually in the area of the cores and cavities. Some of the more common enhancements are mentioned in the next section, along with the advantages of each.

Tool Surface Treatments

As plastic injection molding technology pushes us into the 21st century, we find that major improvements are being made in machines and materials, but one area we are now beginning to look at closely is tooling. Greater requirements are being placed on the length of time a mold is expected to run before it is considered worn out. Some of the new resin formulations and alloys are tough on existing mold materials. Added strength from reinforcements increase property values of the plastics, but their abrasive nature can wreak havoc on the steel finishes of a mold.

This section is not intended to be all-inclusive, but is written as a reference tool for those interested in knowing about, and understanding, some of the up-to-date tool steel enhancement products and how they fare concerning advertised application methods, advantages, and the mold materials on which they can be used. Many of the specific enhancements are known by various tradenames, not all of which are listed here. We are not endorsing any of these products or methods and believe that users must make that selection for themselves.

Thin-metallic Coatings

Dicronite DL-5™ (Dicronite Dry Lube [818] 967-3729). A modified tungsten disulfide in lamellar form, it is applied at ambient temperature by air delivery and bonds by immediately penetrating the mold steel. It is primarily used instead of mold release agents and can be used on all mold materials and platings.

WS2™ (Micro Surface Corp. [815] 942-4221). Also a modified tungsten disulfide in lamellar form, it is applied at ambient (room) temperature using pressurized air and can be used on all stable metal surfaces. It becomes part of the substrate, taking on the same hardness, and cannot be removed without removing part of the substrate surface.

PTFE-metallic Fused Coatings

Poly-Ond™ (Poly-Plating, Inc. [413] 593-5477). A nickel phosphorous material impregnated with fluorocarbon resins including polytetraflouroethylene, hereafter referred as PTFE (also known as Teflon™), it is applied by electroless nickel deposition, followed by a polymer bath, and then baked at 700° F (371° C) to set the polymer into the surface. It can be applied over steel, aluminum, brass, bronze, cast iron, and most other typical moldmaking materials. It is used to eliminate spray releases, prolong tool life, and increase surface hardness to 70 R_c.

TFE-Lok™ (Forestek Plating Co. [216] 421-2552). This is a PTFE-impregnated, hard chrome material that is applied by electrodepositing the chrome, then heating the chrome to expand the pores that are then impregnated with iced TFE particles under high pressure. The process is performed at 400° F (204° C) and can be placed on any steel, stainless steel, aluminum, copper, or copper alloy mold material. It is used for its permanent release properties and increased hardness to 70 R_c.

Nedox-SF2™ (General Magnaplate of Texas [817] 649-8989). A modified electroless nickel infused with PTFE polymer, the nickel is modified to increase porosity, and the polymer is typically applied and set by heat at 750° F (399° C). It is used to precisely control thickness of the mold surface for close-tolerance parts, and can be used over all ferrous materials and some nonferrous, including aluminum. It imparts a hardness up to 70 R_c.

Nituff™ (Nimet Industries [219] 287-7239). A PTFE coating applied over a hard-coat anodizing that is only used over aluminum mold surfaces. The surface is first anodized, then dipped in PTFE at approximately 200° F (93° C). It increases flow and release characteristics and increases wear resistence by imparting a hardness of up to 62 R_c.

PTFE-metallic Codepositions

Nicklon™ (Micro Surface Corp. [815] 942-4221). A 10.5% phosphorous-nickel alloy with 25% PTFE suspended in the matrix, it is applied at 150° F [66° C) by codepositing the nickel and PTFE, and can be used on any metal substance. It is used for its mold release properties and, if heat treated, can reach a hardness level of 70 R_c.

Nicotef™ (Nimet Industries [219] 287-7239). A combination of PTFE submicron particles suspended in a nickel-phosphorous matrix, it is applied through a codeposition process at 195° F (91° C) and can be used on most metals, including aluminum. In thin coatings it is used for release, and in thicker coatings for corrosion resistance. It can attain a hardness of 35 R_c as applied, or 46 R_c if heat treated.

NYE-TEF™ (Electro Coatings of Texas [713] 923-5935). Also a suspension of submicron PTFE particles suspended in a nickel-phosphorous matrix, it is applied by an autocatalytic codeposition process at 95° F (35° C) over most metal surfaces, including brass and aluminum. It is used for its release properties and anticorrosion and can reach a hardness level of 48 R_c as applied and 68 R_c if heat treated.

Nutmeg Chromium-plus™ (Nutmeg Chrome Corp. [860] 953-5411). A chrome/PTFE combination with the PTFE intermixed rather than baked on, it is applied through electrolytic deposition at 170° F (77° C) over steel, stainless steel, or brass, but not aluminum. It is used primarily to resist abrasion and can attain Taber abrasion resistance readings 24% greater than chrome alone. It has a hardness rating of 70 R_c.

Metallic Platings

Industrial hard chrome. This is considered an industry standard for adding thickness to worn surfaces by plating. It is available from many suppliers (contact the American Electroplaters Society at [407] 281-6441) and is applied by electrodeposition of chromium plus trace amounts of oxides and hydrogen at 140° F (60° C). It has a hardness level of 70 R_c and good adhesion on most metals, but softens at 400° F (204° C).

Electrolizing (Electrolizing, Inc. [401] 861-5900). A proprietary process using a nonmagnetic high-chrome alloy in a deposition process at 90° F (32° C), it can be used on any tool steel, 4100 steel, stainless steel, or aluminum. It increases wear resistance, reduces friction, aids mold release, and can be used for repair. It has a hardness level of 72 R_c.

Armoloy™ (Armoloy of Illinois [815] 758-6691). A dense chrome alloy that is applied to ferrous and nonferrous metals, except for aluminum, in an electrocoating process at 140° F (60° C), it has good release and wear properties, and is excellent for use with glass-filled materials, with a hardness rating of 72 R_c.

Electroless nickel. This is the most popular deposited surface enhancement for molds. It is a nickel alloy combined with varying phosphorous content to provide specific properties. Applied through electroless deposition at 180° F (82° C) and up to 400° F (204° C) in postbake, it attains a hardness rating of 48 R_c as applied, and 70 R_c when heat treated. Suppliers can be found by contacting the American Electroplaters Society at (407) 281-6441.

Nutmeg tungsten nickel (Nutmeg Chrome Corporation, [860] 953-5411). A solution alloy of tungsten and nickel, it is applied using electrolytic deposition on any metallic surface at 150° F (66° C) and attains a hardness rating of up to 65 R_c, if baked. It is excellent for corrosion resistance and polishes well.

Surface Hardening Treatments

Melonite™ (Kolene Corp. [800] 521-4182). A thermomechanical, salt-bath nitriding process that reacts to any ferrous based metal substrate to create nitrides, it is performed at 1,075° F (579° C) and achieves up to a 70 R_c hardness. It is extremely resistant to wear and improves fatigue properties by 20 to 100%.

Ion nitriding (Sun Steel Treating [810] 471-0844). A plasma glow surface hardness treatment process in which ionized nitrogen electrons form nitride compound and diffusion zones in most ferrous substrates, it is performed at 950° F (510° C) and achieves a typical hardness rating of 65 R_c. This is a very low cost process used to improve fatigue resistance, wear resistance, and lubricity.

Thin Film, Hi-hard Coatings

Titanium nitriding. This is a common high temperature coating applied through a negative ionization, vapor deposition process at 950° F (510° C). It is applied to most steels, stainless steel, and beryllium copper, and is used primarily to reduce friction. It can achieve a hardness rating of 85 R_c.

Diamond Black™ (Diamond Black Technologies [800] 368-9968). Reported to attain the highest hardness rating available at 95 R_c, it is a thin boron carbide film combined with tungsten disulfide applied in a low temperature (250° F [121° C]) magnetron sputtering process. It is used to increase hardness for extended tool life and also provides good lubricity and corrosion resistance.

Impregnated Polymer

Micro-tuff™ (Plating Resources, Inc. [216] 963-6360). This is a process consisting of impregnating two long-chain polymers into a preplated surface of chrome over most metals, at a temperature that is only referred to as being above ambient (room temperature). It attains the same hardness of the substrate, and is used primarily to improve mold release properties, but can increase wear resistance by 10 times more than that of chrome.

For more detailed information, relative costs, and unlisted advantages and disadvantages, please contact the suppliers shown earlier for these most popular treatments.

Product Surface Finish and Special Requirements

For most applications, the surface appearance of the molded product is important, either from an aesthetic viewpoint or from the standpoint of applying additional finishes, such as a hot-stamped company logo or decorative artwork. The condition and appearance of the molded surface is directly dependent upon the condition of the surface of the cavity in the mold and the parameters of the injection process. While the processing parameters cannot be controlled by the mold designer (or moldmaker), it is necessary for the mold designer to define molded product appearance requirements so that the mold can be fabricated with the proper surface condition.

Each type of machining process results in a roughness of the mold surface, which must be understood for proper finishing of that surface. For example, using EDM to create a cavity surface results in a finish that is considered standard, or between 10 and 100 μin. (250 and 2,500 μm) of roughness. The higher the reading, the rougher the surface. A mirror finish, or lens, may require a reading of less than 1 μin. (25 μm). An average reading for most parts might be in the area of 20 to 40 μin. (500 to 1,000 μm).

Generally speaking, the smoother the cavity finish, the easier it is to eject the molded part. This can be achieved by minimizing hills and valleys that are created during standard machining methods. These are shown in Figure 2-4. The heavy line in Figure 2-4 shows the degree of roughness of the cavity surface. In the top half of Figure 2-4, a typical finish is shown after an EDM machining process. After the proper stoning and polishing process, the roughness has been reduced to that seen in the bottom half of the drawing. The plastic material will be easier to eject from a surface that is smoother, especially if it is a sidewall.

The Society of the Plastics Industry has created a set of standards associated with mold cavity surface finishes. These range from A-1 for the smoothest possible finish, to D-3 for a coarse, dry-blasted finish. There are three levels (1, 2, and 3) in each of the four grades of finish (A, B, C, and D). The SPI has made available a kit showing examples of all these finishes as they appear on metal samples. The A-1 is equivalent to a finish of 0 to 1 μin. (0 to 25 μm), with B-3 being equivalent to a finish of 180 μin. (4,500 μm) and the other grades falling in between these numbers.

It is possible to create a surface that may be too smooth. Depending on the plastic being molded (with hard, stiff resins being the biggest offenders), an ultra-smooth finish may actually create a vacuum between the metal surface and the plastic material, especially when the plastic shrinks on to a core. In that case, the metal surface of the core can be roughened slightly to minimize the vacuum condition. Or, if the product and mold design allow, a poppet valve system can be

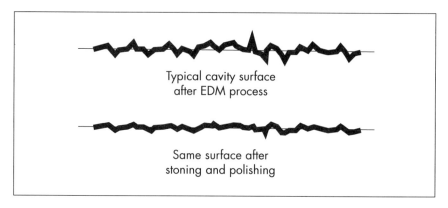

Typical cavity surface
after EDM process

Same surface after
stoning and polishing

Figure 2-4. Microsurface of a molding cavity.

employed to inject a blast of air through the core. This will eliminate the vacuum and allow the plastic part to be ejected.

If the molded part gets a finish applied after molding, the surface will usually require some form of preparation. For example, in the case of polyolefins, the surface is oxidized to accept any type of paint, dye, hot-stamp, or other decorative finish. It is necessary to eliminate the need for mold release agents during the injection molding process whenever possible because they will interfere with the adhesion process. This is another reason for utilizing a high level of polished finish on the mold surface. It will aid in ejection and minimize the need for mold release agents. Any surface that will be decorated during a postmold operation should be identified on the product drawing. This alerts the moldmaker and molder to the fact that those surfaces are critical and require special attention.

Gate Method and Location

The final quality, appearance, and physical properties of the molded product will depend upon where the gate is located and what type of gating system is used. Ideally, the cavity should be gated so that the material will enter the thickest section of the part first, as shown in Figure 2-5.

The basic concept for this is based on the fact that the molten plastic molecules will tend to take up the space available to them. Also, they will try to have equal air space between them. If the gate is located in the thickest section of the cavity, the molecules will be forced together and compacted as they travel into the cavity. The air between them will be forced out. This results in a very compact molecular structure that yields a molded part with the greatest degree of structural integrity possible.

If the same part were gated at the thin end, the molecules would be allowed to expand as they flowed through the cavity, and the air spaces between the mol-

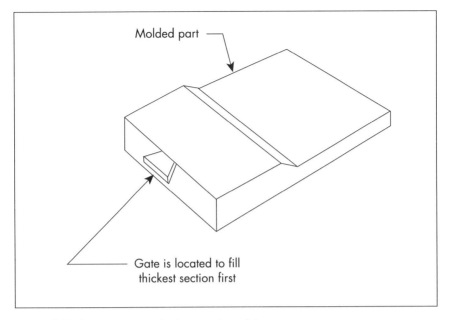

Figure 2-5. Gating into the thickest section of the part.

ecules would grow, resulting in a weaker molecular bond and a molded part with low structural integrity.

Ideal gate location and design will be covered in a later chapter. The primary focus of activity right now is to be aware of potential gate locations so steps can be taken to notify the product designer of any problems. A gate of any type will leave evidence (called a *vestige*) of its existence. Depending on the type of gate used, that vestige will be left sticking out of the molded part, or broken into the molded part. It will never be perfectly flush with the molded part. If the vestige will interfere with the action, appearance, or intended use of the molded part, the gate may have to be relocated. The product designer must be involved in that decision.

Ejector Method and Location

After the molten plastic solidifies in the mold, the final molded product must be ejected from the mold. The most common method uses ejector pins. These pins are used to push the molded part out of the cavity that formed the part, as shown in Figure 2-6.

Ejector pins should be as large in diameter as possible to distribute the ejection force over the molded part. This will minimize stress due to ejection and aid in minimizing cracks or punctures caused by too small an ejector area. Ejector pins

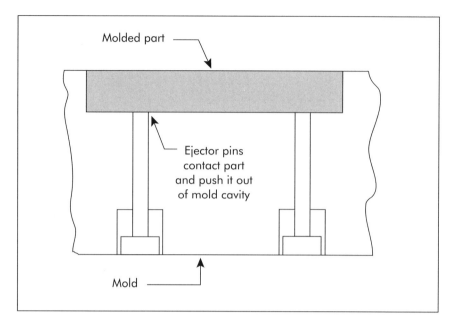

Figure 2-6. Ejecting molded part from mold.

should be located where they are pushing on the strongest area of the part, if possible. This would be near corners, under bosses, and close to rib intersections. Round ejector pins are the most common and the least expensive to make, but rectangular cross sections also can be used.

As with gates, ejector pins will leave evidence of their existence on the molded part. Due to constant expansion and contraction of the various mold components during molding, ejectors can not be made to stay perfectly flush with the surface of the molded part. Therefore, they leave a pad of excessive plastic (protrusion) if the pins are too short, and they leave an impression in the plastic part if they are too long. This evidence of existence is called a *witness mark*, as shown in Figure 2-7.

If the pins are long, thus creating an impression in the part, the molded part will tend to stay on the ejector pins and may not fall out of the mold when ejected. This can cause considerable damage if the mold were to close on the non-ejected part. Therefore, it is usually wise to make the pins intentionally short. This will result in a thin pad of excess material. The product designer must be made aware of where ejector pins are intended and what the witness mark will consist of to make a decision of acceptance. If the intended marks are not acceptable (due to part function or appearance), a different type of ejection (such as a stripper plate or an expensive air blast system) may need to be employed. Or, the part may have to be situated differently in the mold to allow ejector pins to be relocated. This, too, will result in higher mold costs.

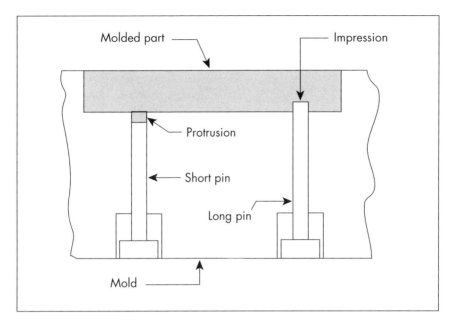

Figure 2-7. Ejector pin witness marks.

Location of Cavities and Cooling Channels

If a single cavity mold is being used, the cavity should be placed directly in the center of the mold. This allows for sprue gating and creates the best conditions for molding. The material injects immediately into the cavity and travels the least distance. There are no restrictions, so injection pressure can be minimized and stress is greatly reduced. These are the conditions we must attempt to duplicate even in a multicavity mold.

In multicavity molds, we must place the cavities as close to the center of the mold as possible, but keep in mind that we need ejector pins for the parts as well as for the runners used to bring material to the cavities. In addition, we must place cooling channels in the mold plates to bring coolant (usually water) as close as possible to the mold cavities without breaking through the steel and causing water leaks. Be careful to locate the cavities so that mounting bolts and ejector pins are not in the way of the cooling channels. As the number of cavities increases, the layout becomes more complicated and the process becomes more difficult. A good rule-of-thumb is that cooling channels should be located no closer than twice their diameter from any other object, as shown in Figure 2-8. This allows for enough metal surrounding the channel to minimize breakthrough tendencies.

An ideal layout for a multicavity mold would be like spokes in a wheel. The spoke layout allows cavities to be as close as possible to the center of the mold

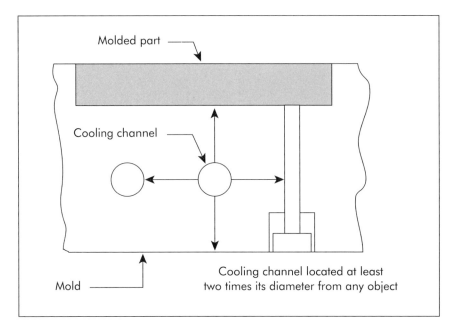

Figure 2-8. Cooling channel spacing.

and eliminates any right angle turns in the runner system. These turns result in a 20% pressure drop for each turn, which creates the need for increasing the diameter of the runner to ensure proper material flow. This increase results in higher material costs and longer cycle times and should be avoided if possible. Figure 2-9 shows a typical spoke layout for an eight-cavity mold.

One disadvantage to the spoke concept is the total number of cavities possible in a given size of mold. A squared pattern can accommodate many more cavities than a spoke pattern, as shown in Figure 2-10. However, the squared pattern requires turns in the runner system. Commonly, these turns are represented as right angles. But, every time a runner makes a right angle turn, it requires additional injection pressure to push material through the turn. Therefore, in an effort to keep pressures equal at all times, the primary runner diameter must be increased by 20% to accommodate the pressure drop. If squared patterns are necessary, it is better to have runners with sweeping turns instead of right angles, as shown in Figure 2-11.

Remember that no matter which runner system is used, ejector pins must be utilized to eject the runner system as well as the molded part. Therefore, the cavity layout must be concerned with not only how close to place the cavities to the center of the mold for minimal material travel, but also how to avoid ejector pins (and mounting bolts) being located in the middle of cooling channels.

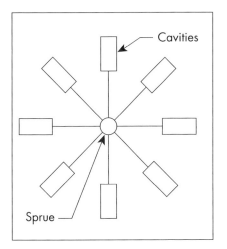

Figure 2-9. The spoke concept for cavity layout.

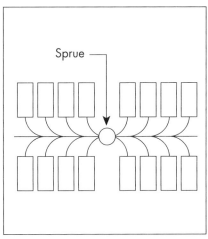

Figure 2-11. A sweeping turn runner system.

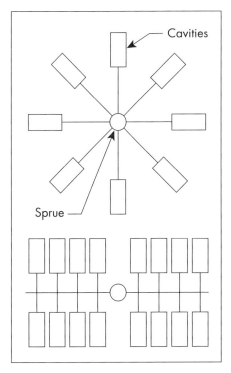

Figure 2-10. A squared versus a spoke layout.

Runner Cross Section

The ideal runner cross section is round. This results in equal injection pressure in all directions, which means there is no cause of stress. If a cross section other than round is used, injection pressures are unequal and stress will result. This is caused by deformation of the plastic molecules as they travel through such a cross section, as shown in Figure 2-12.

The arrows represent the direction of pressure as applied by the injection unit to the material as it is pushed along the runner through the mold. As can be seen in the upper left drawing of Figure 2-12, the runner with the round cross section has pressure applied equally in all directions. This is shown by the arrow lines all being of equal length. The upper right drawing shows how the pressure is applied through a square (or rectangular) cross

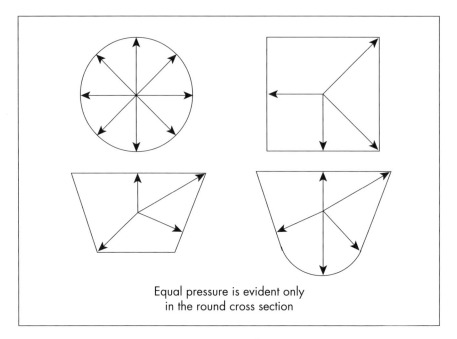

Equal pressure is evident only
in the round cross section

Figure 2-12. Various runner cross sections and stresses.

section. Note the varying lengths of the arrow lines. This demonstrates that pressure varies depending on exact locations in the runner. Unequal arrow lines indicate unequal pressures, and unequal pressures result in stress being created in the plastic material. The stress is caused by molecules being squeezed and/or expanded due to their natural tendency to take up the space that is available to them.

The stress conditions get worse if we attempt to utilize a cross section that is trapezoidal in shape, as shown in the two bottom drawings of Figure 2-12. Here, the arrow lines exist in a multitude of varying lengths and the stress conditions are multiplied accordingly.

This is not to say that acceptable parts can be molded with only a round runner system. It only indicates that the amount of stress molded into a finished product is greatly increased by a cross sectional shape of the runner system other than round. The specific design requirements of any product may be flexible enough to allow stress that is additional to the normal stress introduced by the molding process itself. If that is the case, runner systems other than round may allow the molder to get by and produce an acceptable product. However, due to the fact that any molded-in stress will eventually be released (it could take a few weeks or many years), it is better to mold with as little stress as possible. Therefore, a round runner should be used whenever possible.

Of course, what we have been discussing is surface runner systems. But, even if hot runner, or insulated runner, or three-plate runner systems are employed, the round cross section is still the preferred shape because the pressure is applied equally in all directions.

Venting Concepts

When an injection mold is closed up in preparation for receiving molten plastic, the area that forms the image to be molded (the cavity image) has a quantity of air trapped in it. This air will compress as the plastic is injected, but that takes pressure. In fact, much of the pressure needed to fill a mold with plastic is needed to push out the air that is trapped in the cavity image. The air will compress until it becomes so dense it ignites and burns the surrounding plastic. This results in a charred appearance to the plastic and a nonfill condition on the part in the area where the air was compressed.

To minimize the amount of injection pressure needed to fill the mold, and to eliminate the burning of trapped air, the trapped air is removed from the mold by *venting*. Venting usually consists of grinding thin flats on the parting line surface of the mold running from the edge of the cavity image to the edge of the mold and thus to the atmosphere, as shown in Figure 2-13.

As the trapped air and any volatile gases from the melt are pushed out of the mold cavity, it is easier for the incoming molten plastic to enter and fill the mold completely. The faster the air is eliminated, the quicker and easier the plastic can enter. If a mold is properly vented, it is possible to reduce the injection pressure by as much as 50%, thus reducing the amount of molded-in stress. In addition, it costs less to prepare the melt because we do not need as much injection pressure resource or as high a temperature on the melt to make it fill the cavity image. And, we can reduce the total cycle time by a second or more because the melt fills so quickly. As you can see, proper venting is desirable for reducing product cost, improving part quality, and reducing stress and rejects. The addition of a vacuum system can improve the venting process.

As part of the mold designer's responsibility, venting should be discussed with the moldmaker before the mold is built. This will ensure that proper venting is in place even before the first part is molded, and this in turn will help produce acceptable parts quicker and with less hassle than with a nonvented mold or one in which vents are only added after molded parts show burn marks caused by trapped air.

Dimensioning the Mold Design

After the initial, basic mold layout has been determined, with ejector pins located and runners and gates in place, the designer is ready to indicate dimensions for all details of the mold. The proper way to do this is to work with the moldmaker to decide which equipment will be used to manufacture each of the various details.

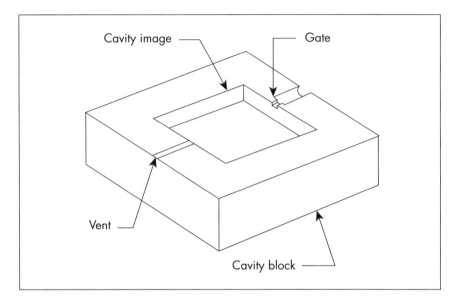

Figure 2-13. Typical venting concept.

Then, the designer dimensions those details in ways that are most economical for the chosen fabrication method.

A starting point must be established for all dimensions related to cavity and core placement and fabrication. This starting point should be determined between the mold designer and moldmaker and will be utilized as a point from which all vital dimensions will radiate, as shown in Figure 2-14.

At this point, the designer and moldmaker should discuss those details that are known to need replacement (such as gate inserts). These details should be fabricated in a way that ensures interchangeability with a minimum of final fitting required. If fitting is required, the original mold design must plan for easy fitting at a later date.

A question to ask during these proceedings is: How will these details be fabricated, and what information will be needed to accomplish the task? The answers will assist the designer in dimensioning the details so that the moldmaker can create the details with the least amount of confusion and use the most economical method.

MOLD LIFE EXPECTANCIES

How long should a mold last? Five years? Ten years? One year? The answer depends on how many cycles are needed to produce the total number of parts required for the life of the program. That can range from one day to decades. The

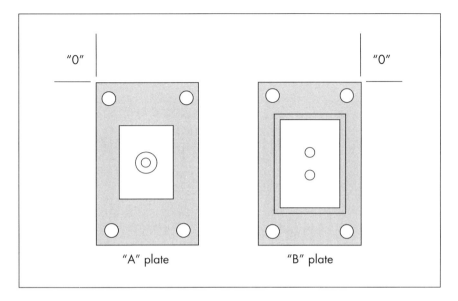

Figure 2-14. Starting point for dimensioning.

longer the mold is expected to survive, the greater the amount of quality that must go into designing and building it. This usually results in higher costs and longer lead times.

Theoretically speaking, a mold can last forever. Just like an automobile, parts can be replaced as they wear out, and improvements can be implemented as they become available. But, also like the automobile, it may become very expensive and time-consuming to try to keep a mold going, and it may be more economical to build a new one.

The following paragraphs discuss some methods of producing plastic parts, based on total quantity required. As can be seen, as the quantity increases, the process to produce the parts becomes more complex, and the tooling becomes more expensive, but the cost of the part drops.

Rapid Systems

There are many new methods of producing low-quantity requirements based on using light-curable resins and/or fast-curing compounds. These are usually formed layer upon layer, very rapidly, to produce a product similar to the final requirement. Most systems use ultraviolet (UV) lasers to provide the light curing source. The finished product is limited to the quality properties of the resin system used. However, new materials are being developed continuously to overcome these limitations. Rapid systems are expensive processes to use, but they are capable of producing a finished product in a matter of a few hours after the design is formalized. They should be used when the required number of parts is minimal (approximately two or three units) and time is critical.

Fabrication

In this process, plastic sheet and rod stocks are machined and bonded together to create a facsimile of the required product. This is one of the oldest methods of producing low quantities and is ideal when time is not critical and cost must be held to a minimum. It is cost-effective for up to five or six units.

Thermoforming

In this process, heated plastic sheets are vacuum (or pressure) formed over models of wood, plaster, latex rubber, aluminum, or other easily machined material, and then cut to shape and bonded to form the finished product. This method requires very little tool investment, but the models are only good for approximately two dozen finished units. The cost of postmachining and bonding must be calculated into the final cost.

Casting

In the casting process, a plastic such as liquid urethane is poured into an epoxy, aluminum, or similar material mold. The two-part casting resin then hardens and is removed from the mold and finish-machined to exact dimensional specifications. This is a good system to consider for up to 25 or 30 units from each mold.

Injection Molding

In this process, molten plastic resin is injected into a closed mold. The resin is selected from a wide choice of materials, depending on the expected life of product and quality required. These molds can be made from aluminum, epoxy, soft steel, prehard steel, tool steel, stainless steel, copper, and combinations of any or all of these materials. Injection molding is considered cost-effective when the required annual quantities are more than 1,000 units. However, fewer units can be justified if the part is complicated and would require many manufacturing operations if it were not injection molded.

Mold life is directly proportional to the degree of quality and sophistication that is used for building it. Some moldmakers are specialists in building high-quality, sophisticated molds, while others are specialists in building molds that do not require such sophistication or quality. Moldmakers must understand their own limitations and not attempt to build molds that are not within the realm of their expertise. For example, why spend a lot of time and money for an expert moldmaker that can build sophisticated molds for electronic connectors, if all you want built is a simple mold to produce small, throwaway flower pots?

OWNERSHIP OF THE MOLD (AND OTHER TOOLS)

In most cases, the person (or company) paying for the mold, and any related secondary tooling, retains ownership of the mold and tooling. In some cases, the molding vendor may utilize a universal-type mold base that accepts only cavity inserts built for the customer. In that case, the customer only retains ownership of the individual cavity inserts. If the customer wanted to move the job to another molding vendor, it would be necessary to locate one having the same type of universal base. Of course, another possibility would be to convert the individual cavity inserts into stand-alone molds, or to fit another universal-style base.

While the cost of the mold and tooling is usually paid for up front by the customer, the molding vendor does have a responsibility to maintain and keep the customer's tooling in reasonable condition, determined by the quantity of parts being produced and expected life of the tooling. This type of maintenance would include such items as replacing broken ejector pins, use of a rust-preventive during storage periods, and periodic cleaning and polishing of molding surfaces. A general rule-of-thumb is that the vendor accepts responsibility for maintaining the tooling after it is capable of producing acceptable parts. The customer must provide tooling capable of producing those good parts. In some cases, the molding vendor is also the moldmaking vendor, and thus takes full responsibility for the building of the mold and molding of the parts. In most cases, the maintenance of the tooling will result in annual expenditures of approximately 7% of the initial cost to build the mold. Therefore, if a mold cost $30,000 to build, the molder should expect to pay approximately $2,100 a year to maintain it in as-new condition. This includes labor and part costs, as well as the cost of materials, such as rust preventives and mold-cleaner spray. In addition, a molder who causes damage to the mold for any reason is responsible to repair that damage and must bear the expense.

In most states, if a customer does not finish paying monies due, the molder immediately assumes a mechanic's lien on the mold and may sell it to recover payment.

SUMMARY

Before any work is actually done to create a mold design, there are some basic facts that need to be gathered. These include determining how many cavities to build, what material to make the mold out of, and other data.

While the mold designer is not normally the one to select which plastic material is to be molded, the designer should be aware of some of the more important aspects and characteristics involved in molding specific plastics. For example, shrinkage factors sometimes vary widely between different materials and may vary even among different grades and versions of the same material. Also, some plastics will absorb and dissipate heat more efficiently than others, resulting in more efficient cooling during the molding process. This may affect cooling channel locations in the mold. And the viscosity of a particular plastic has a large bearing on gate, runner, and vent design, location, and construction.

Every material we know of (except water) expands when it is heated and contracts when it is cooled. In the field of plastics we define the contraction phase as *shrinkage*. Each plastic material has *a shrinkage factor* assigned to it. This factor is used to estimate how much a part will shrink after it is removed from the mold. After that is determined, the mold can be built to a set of dimensions that creates a molded part large enough so that it will contract to the desired finished size after shrinkage.

The viscosity of a plastic determines how much pressure will be needed to inject the material into a mold. Viscosity is measured by way of elaborate and relatively expensive test equipment. But, it can be indicated inexpensively by using ASTM test D1238 that uses a small amount of plastic material and simulates the injection molding process. This test is called the *melt index* test, but is also known as *melt flow, flow index*, and *melt rate*. A machine called a *plastometer* is programmed to a set of conditions dictated by the plastic being analyzed.

Before we can determine the size of mold and the size of equipment needed to run the mold, we must determine how many cavities are required. Along with the total time of a cycle, the number of cavities determines how many molded parts can be produced during one complete cycle of the injection molding process. The number of cavities needed depends on the time frame established for producing the annual volume requirements of a specific product.

The final quality, appearance, and physical properties of the molded product depends on where the gate is located and what type of gating system is used. Ideally, the cavity should be gated so that the material will enter the thickest section of the part first.

As with gates, ejector pins will leave evidence of their existence on the molded part. Due to constant expansion and contraction of the various mold components during molding, ejectors can not be made to stay perfectly flush with the surface of the molded part. Therefore, they leave a pad of excessive plastic if the pins are too short, and they leave an impression in the plastic part if they are too long. This evidence of existence is called a *witness mark*.

An ideal layout for a multicavity mold would be like spokes in a wheel. The spoke layout allows cavities to be as close as possible to the center of the mold and eliminates any right angle turns in the runner system. These turns result in a 20% pressure drop for each turn, which creates the need for increasing the diameter of the runner to ensure proper material flow. This increase results in higher material costs and longer cycle times and should be eliminated if possible.

To minimize the amount of injection pressure needed to fill the mold, and to eliminate the burning of trapped air, we remove the trapped air from the mold through a process called *venting*. Venting usually consists of grinding thin flats on the parting line surface of the mold, running from the edge of the cavity image to the edge of the mold, and thus to the atmosphere.

In most cases, the person (or company) paying for the mold, and any related secondary tooling, retains ownership of the mold and tooling. While the cost of the mold and tooling is usually paid for up front by the customer, the molding vendor does have a responsibility to maintain and keep the customer's tooling in reasonable condition, determined by the quantity of parts being produced and expected life of the tooling. In most states, if a customer does not finish paying monies due, the molder immediately assumes a mechanic's lien on the mold and may sell it to recover payment.

QUESTIONS

1. Which category of plastics tends to shrink more in the direction of flow rather than across the direction of flow?
2. Define the different values for low, medium, and high shrinkage factors.
3. What is the name and number of the ASTM test used to indicate viscosity?
4. Why do we need to determine how many cavities will be in the mold?
5. What is the most common steel used for making support pillars?
6. What are the SPI number designations for both the smoothest cavity finish and for a coarse, dry-blasted finish?
7. Why might an ultra-smooth finish cause molding problems?
8. What problem can occur if ejector pins are too long?
9. What is the ideal runner layout for a multicavity mold?
10. What is the ideal runner cross-section shape, and why?
11. How do we remove trapped air and gases from the mold cavity?
12. How many years can a mold be expected to last?

Basics of Mold Construction 3

DEFORMATION TENDENCIES

A mold is made up of a variety of plates, blocks, and other shapes of metal (or similar material). Through the application of injection pressure and clamp tonnage, these components are exposed to tremendous forces that tend to twist, compress, warp, and bow them. This tendency toward deformation must be controlled by properly designing and locating the components that make up the finished mold. The following section addresses some of the fundamental methods of achieving this for cavities, cores, and shut-off lands, cavity retainer plates, waterline locations, support plates and support pillars, and ejector housings.

Cavities, Cores, and Shut-off Lands

This section will discuss cut-in-the-solid sets, free standing sets, pocketed cavity sets, and laminated construction as it relates to producing cavities, cores, and shut-off lands.

Cut-in-the-solid Sets

The simplest method of machining a cavity image is to place it directly into one of the main plates of the mold. These are called the A and B plates and they serve as the plane at which the mold separates when opening. While the combination of core and cavity make up what is called the *cavity set*, a core is actually any component with a shape that is convex or protruding (also called a *male component*) and a cavity is any shape that is concave or shaped as a depression (also called a *female component*). The *cavity image* is another term for the cavity set. Figure 3-1 shows a cavity image that can produce a part.

If the cavity set is machined directly into the A or B plate, as shown in Figure 3-2, it is said to be *cut-in-the-solid*. This does not lend itself well to corrections, repairs, or engineering changes, but is an inexpensive way of producing a cavity image and can be cost-effective for short runs, or when quality or dimensional control is not critical.

Note in Figure 3-2 that there is a raised wall around the cavity image. This is approximately .002 to .003 in. (0.05 to 0.76 mm) and is called the *shut-off land area*. This is created by machining away the metal from the surface of the A and B retainer plates to within approximately .5 in. (13 mm) of the cavity image. The resultant land acts as an impact area and concentrates all the clamp force around the

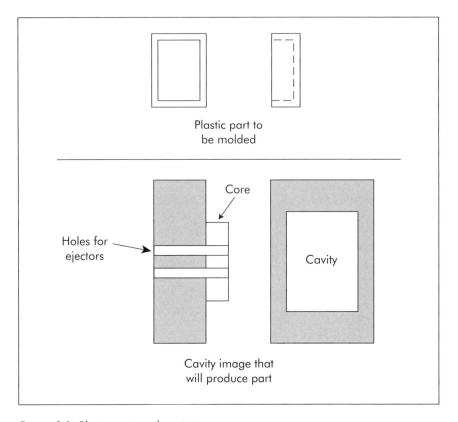

Plastic part to
be molded

Core

Holes for
ejectors

Cavity

Cavity image that
will produce part

Figure 3-1. Plastic part and cavity image.

shape of the cavity image. Without the shut-off land, the clamp force would be distributed over the entire face of the A and B plate, which would result in requiring higher clamp pressures to keep the mold closed against injection pressure.

The biggest disadvantage of cutting in the solid is that it is practically impossible to properly vent any deep areas of the image. This will result in the need for extreme injection and clamp pressure and cause excessive stress to be molded into the part.

The second biggest disadvantage of cutting in the solid is the difficulty in making repairs or engineering changes to the cavity image. Laminated sections are much easier to remove, manipulate, machine, and transport than an entire solid-cavity block.

Free-standing Sets

Another inexpensive method of producing a cavity set is to place it within a block which is then attached to the very surface of the A and B plates. This is referred to

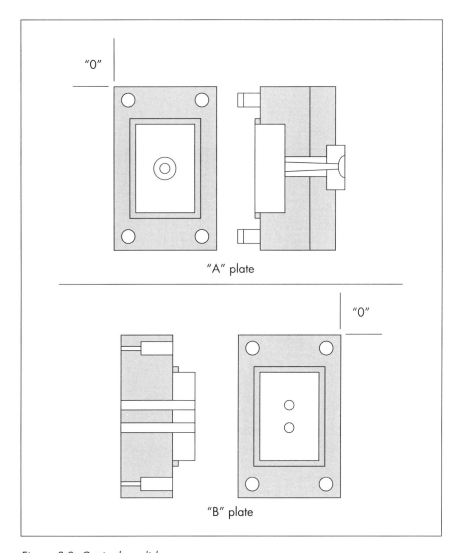

Figure 3-2. Cut in the solid.

as a *free-standing cavity block* and is the least expensive of the methods discussed in this book. Figure 3-3 shows the concept behind this method.

The free-standing cavity block mounts directly to the A and/or B plate and the
mold closes against the mating faces of the free-standing blocks. This becomes
the shut-off area.

It is important to realize the requirement for sidewall strength of the block.
There is nothing to retain them, so they must provide their own strength. This is

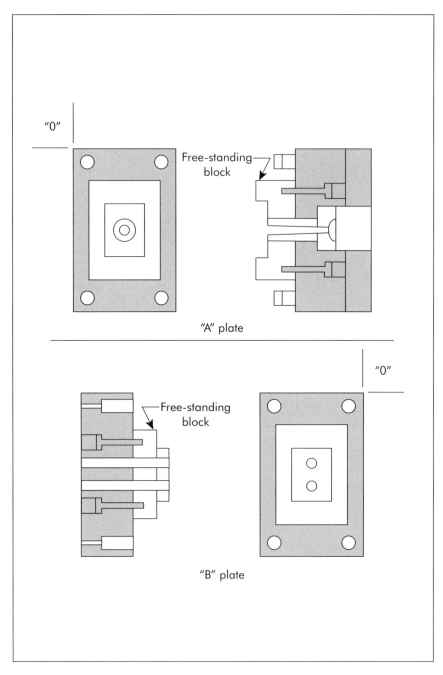

Figure 3-3. Free-standing cavity block.

done by ensuring that the sidewalls have enough steel in them by employing a rule-of-thumb that requires a minimum sidewall thickness of 1.5 times the depth of the part to be molded. For example, in Figure 3-3, the area above the part depth is 1 in. (25 mm). Each sidewall of the cavity block must then be at least 1.5 in. (38 mm) thick. This will ensure that the sidewall is strong enough to overcome the forces caused by injection pressure of the incoming molten plastic. If the part has an unusual height-to-width ratio (beyond 2 to 1) this factor should be increased accordingly. Any attempt to reduce this deminsion may result in extended cavity damage and injury to personnel.

Pocketed Cavity Sets

It is better, if possible, to machine the cavity sets as separate blocks which are then inserted into pockets that are machined into the A and B plates (which are also referred to as *cavity retainer plates*). When utilizing this method, it is said that we are *pocketing* the cavity sets, as shown in Figure 3-4.

The cavity sets are placed in the pockets and are left standing above the retainer plates approximately .002 to .003 in. (0.05 to 0.08 mm). This creates the shut-off land area. Although it is not a good practice (due to potential loss of loose pieces), it is common to use thin metal shims under the cavity sets to achieve this raised situation to allow for easy adjustment if it is needed. If shims must be used, they should be identified with an engraver to minimize the possibility of mixing shim sizes after re-assembly of a mold due to cleaning or repair.

When using the pocketing method, it is wise to incorporate tapered wedges for securing the cavity set in the pocket as shown in Figure 3-5. This is especially true in utilizing laminated cavity sections as described next. The wedges can be mounted from either the parting line face or the backside face of the A or B plate, but it is easier to remove the wedges (for disassembling the mold) by mounting from the parting line side and using jackscrews to remove the wedge from the backside, as shown in Figure 3-6.

Laminated Construction

If a mold designer or moldmaker decides to use the pocket method mentioned earlier, the cavity sets themselves can be made as either a solid piece, or laminated with individual sections mounted together. The latter is called *laminar construction* and is the preferred method. As shown in Figure 3-7, the laminar method (although more expensive) allows for ease of repair, closer dimensional control because all surfaces can be ground, and creates areas to place vents for deep sections or complicated surfaces where air can be trapped.

Cavity Retainer Plates

The cavity retainer plates must be thick enough to contain the cavity image without creating the potential of breakthrough on the bottom face, and have enough mass surrounding the cavity image to contain the pressures exerted during injection.

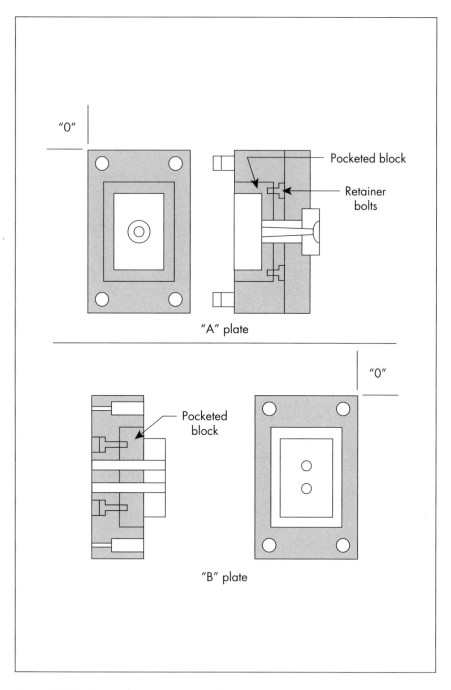

Figure 3-4. Using pockets to retain cavity set.

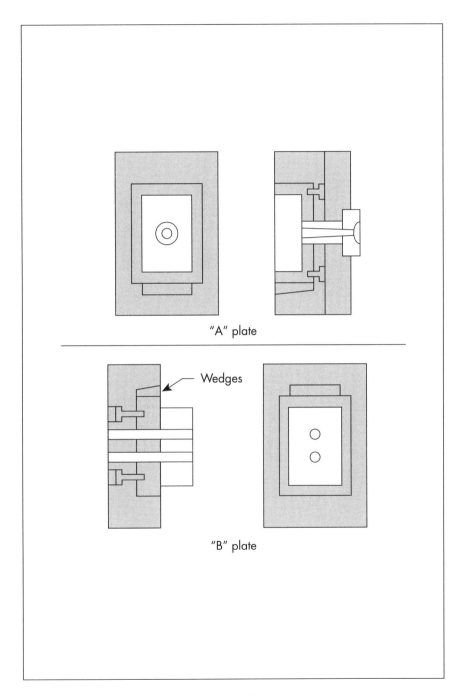

Figure 3-5. Wedges used for locking cavity block.

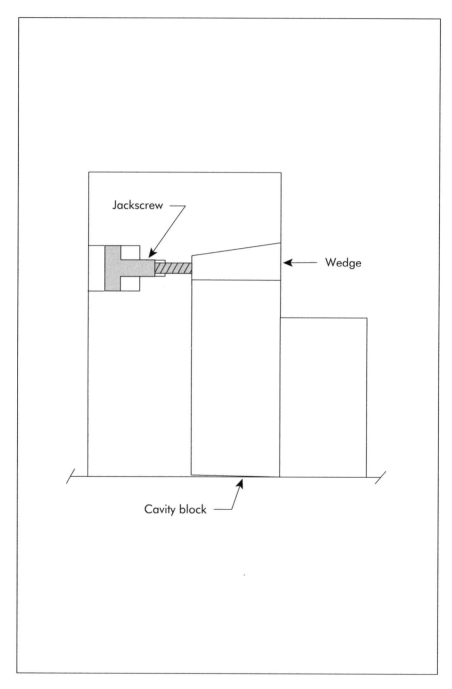

Figure 3-6. Jackscrew to remove wedge blocks.

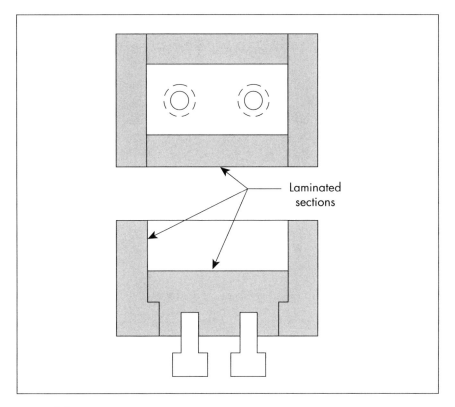

Laminated
sections

Figure 3-7. Laminar cavity set construction.

Proper strength can be achieved by making the sidewall thickness at least 1.5 times the depth of the part being molded, as shown in Figure 3-8. Exceptions can be greater than 1.5, but never less. This holds true for any of the mounting methods we have discussed so far.

The distance between the cavity bottom and the plate (where it is mounted) also should be 1.5 times the depth of the part being molded. Actually, this dimension will usually be determined during design after placing waterlines and mounting bolts in location.

Waterline Locations

The waterlines should be located so there is steel surrounding them to at least two times their actual diameter, as shown in Figure 3-9. The waterlines are usually identified by their nominal "pipe" diameter, which is smaller than the actual dimension. For instance, a 1/8-in. (3-mm) pipe-tap waterline has a drilled hole that is actually 11/32 in. (9 mm) in diameter, a 1/4-in. (6-mm) pipe-tap waterline has a drilled hole that is actually 15/32 in. (12 mm) in diameter, and a 3/8-in. (9.5-

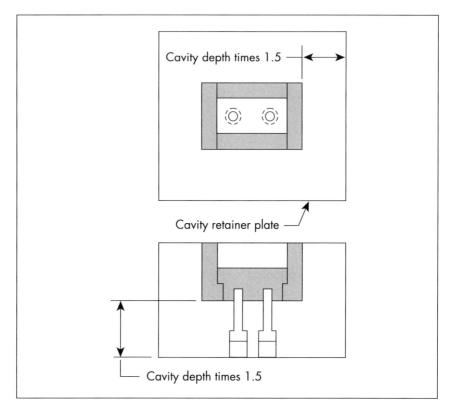

Figure 3-8. Proper steel allowance surrounding cavity set.

mm) pipe-tap waterline has a drilled hole that is actually 19/32-in.(15-mm) diameter. The 1/4-in. (6-mm) pipe-tap waterline is the most common. So, utilizing the rule-of-thumb mentioned earlier (1.5 factor), when using the 1/4-in. (6-mm) pipe-tap waterline, there should be a minimum of 15/16 in. (24 mm), which is twice that of the 15/32-in. (12-mm) diameter between the waterline and any object or the edge of the cavity block, as shown in Figure 3-9.

There are other items to consider when locating waterlines. The primary concern is the placement of ejector pins for removing the molded part and the runner system from the mold. The ejector pins come up through the cavity retainer plates and any backup plates, such as the support plate. Therefore, they could interfere with an ideal location for waterlines, as shown in Figure 3-10. In a situation such as this, the waterline may have to be placed farther away from the cavity than the ideal twice its diameter. If the distance is too great (more than three times the diameter), it may be prudent to increase the diameter of the waterline being used. See the "Waterlines" section in Chapter 6 for more information.

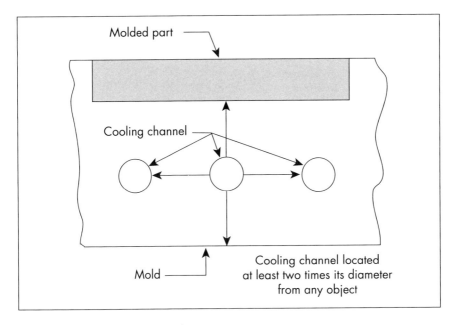

Figure 3-9. Waterline distance from cavity set.

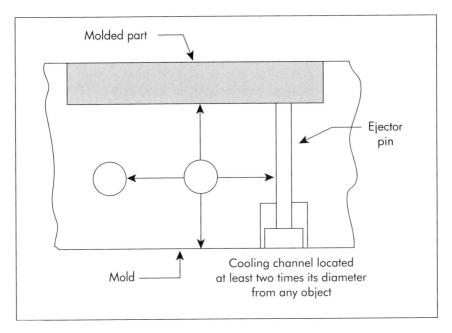

Figure 3-10. Ejector pin interference with waterline.

Support Plates and Support Pillars

When material enters the mold from the molding machine's heating cylinder, it does so at pressures of 10,000 psi (68.9 MPa), on average. This pressure can be as high as 20,000 psi (137.8 MPa), or more in specific circumstances. This pressure enters the sprue bushing of the mold and contacts the face of the B plate at right angles, as shown in Figure 3-11.

The total force of all this pressure (up to 10 tons [89.6 kN] or more) can distort the B plate and cause it to deflect and bow. The bowing may cause plastic to flash around the molded part perimeter and result in a loss of packing pressure needed to fill the part completely. It also reduces the degree of control the processor has over the molding parameter of injection pressure, and it will cause early failure of the B plate. To compensate for this condition, the mold contains two important items: a support plate placed directly behind the B plate, and support pillars placed between the back of the mold and the support plate. These are shown in Figure 3-12.

Support plates are usually 1-7/8- to 2-3/8-in. (48- to 60-mm) thick, with large mold bases requiring thicker support plates. They are supplied to standard nomi-

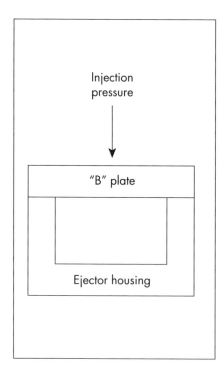

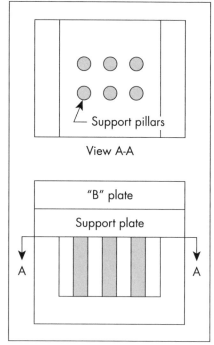

Figure 3-11. Injection pressure directed at B plate.

Figure 3-12. Support plate and support pillars.

nal thickness (by mold base suppliers) based on the length and width dimensions of the A and B plates.

The support pillar dimensions and numbers are determined by the mold designer. They are available in standard diameters of 1, 1-1/4, 1-1/2, 2, 3, and 4 in. (25, 32, 38, 51, 76, and 102 mm). Lengths are from 2-1/2 to 8 in. (64 to 203 mm). Special diameters and lengths are available at special prices and lead times. Note that metric conversions given here are for relative dimensions, not equivalent stock sizes.

The use of support pillars greatly increases the amount of projected area of molded part that a mold can sustain. For example, as shown in Figure 3-13, a mold base that is 11.875 × 15 in. (30 × 38 cm), without support pillars, will permit 14 in.^{2} (90 cm^2) of cavity image (projected area) without deflecting enough to cause flashing. But, by adding a single row of support pillars, the permitted cavity image area is increased by 4 times, to 56 in.^{2} (361 cm^2), as shown in Figure 3-14. And, by utiliz-

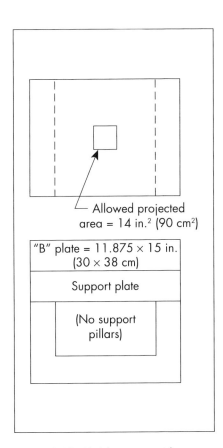

Figure 3-13. Molding area without support pillars.

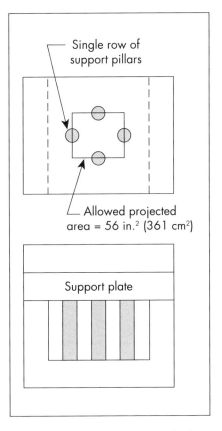

Figure 3-14. Adding one row of pillars.

ing two rows of support pillars, the permitted cavity image area can be increased by 9 times, to 126 in.^{2} (813 cm^2), as shown in Figure 3-15.

The number, size, and location of support pillars is determined by the ejector system layout. The ejector plates must be allowed to slide forward and backward around the existing pillars, which are stationary, as shown in Figure 3-16. Support pillars must be placed and sized so that they do not interfere with the location of ejector pins, which are positioned in the ejector plates. This creates a logistics problem for the designer because the ideal location of a support pillar is directly beneath a cavity image. Unfortunately, that is also the required location for ejector pins. So, the designer must be cautious to provide enough support pillar capacity without interfering with proper ejector pin operation and location.

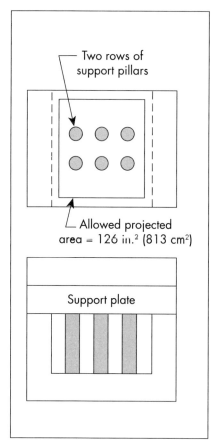

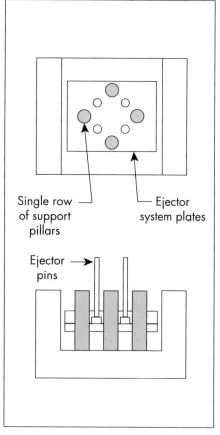

Figure 3-15. Adding two rows of pillars.

Figure 3-16. Ejector plates surrounding support pillars.

Ejector Housings

Ejector housings are U-shaped components of the mold base that house the ejector system. Figure 3-17 shows a standard ejector housing in place.

Figure 3-17 shows a one-piece construction, meaning that the entire housing is machined from a forged solid. The sidewalls are not bolted to the bottom plate, but are integral to it. This provides maximum strength and resistance to the tremendous sidewall forces from the clamping tonnage that keeps the mold closed against incoming injection pressure.

Some ejector housings are made by bolting bolster blocks to a flat plate. While this mimics the one-piece construction method mentioned earlier, it does not provide the same strength and resistance to clamping pressures as the one-piece design. When possible, it is best to utilize the one-piece construction.

CAVITY AND CORE CONSTRUCTION

This section will discuss determining the parting line, shut-off areas, machining methods, other fabrication methods, polishing and finishing, textures, and plating.

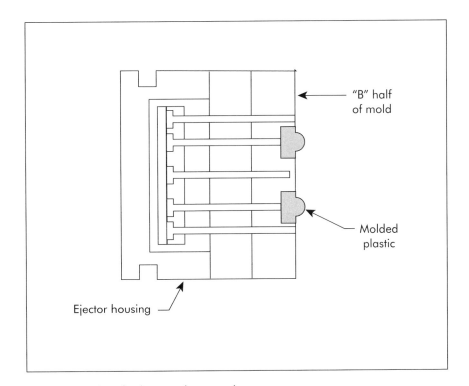

Figure 3-17. Standard ejector housing shape.

Determining the Parting Line

A *parting line* can be defined as any part of the mold that parts, and forms part of the part. There is usually only a single parting line, but there can be two or three or more, depending on the complexity of the mold.

Items that determine the number of parting lines include the geometry of the part to be molded, the number of cavities being produced, the type and style of runner system, the type of gating utilized, and the method of ejection of the molded product.

The primary parting line is always that which divides the A plate from the B plate. This is the plane that is exposed when the mold separates for opening and is shown in Figure 3-18. The parting line follows the contour of the molded part. The more complex the part contour, the more complex (and expensive) it is to create the parting line in the mold.

Ideally, the parting line should be a perfectly straight and level plane, but many part designs require the primary parting line to be stepped and/or contoured. An objective study of the part design will aid in determining the most efficient method and location for the primary parting line. All protruding (male) metal should be on the moving half of the mold, while all concave (female) metal should be on the stationary half of the mold. This is to ensure that, as the plastic cools and shrinks, it will shrink onto the ejection half of the mold, and away from the injection half of the mold. Sometimes, due to special gate design caused by part design, it may be necessary to use a "reverse" parting line. In this case, the male components must be placed on the stationary half of the mold and the female on the moving half of the mold. When this occurs, it is necessary to place an ejection system on the stationary half of the mold because the cooling parts will shrink to that side.

Multiple parting lines occur when part design necessitates special mold design and construction, such as molds that utilize slide mechanisms to create undercuts on a molded part. One such situation would be the case shown in Figure 3-19. Here, a bobbin-shaped part is being molded. The round bobbin has a top face and a bottom face, each of which creates an undercut condition. These undercuts can be formed utilizing a slide mechanism that moves out of the way to allow the finished molded part to be ejected from the open mold. The moving action is caused by slotted holes in the slides following the stationary angled pins. When the mold closes, the slides move in again to create the area that will result in molded undercuts during the next molding cycle. Note that there are actually three parting lines created due to this design. The primary parting line is between the A and B plates, while a second parting line is created on the face of the slides, and the third parting line results from the top and bottom face areas of the slides.

There are situations where our definition of a parting line must be stretched. One such case is with the design of a three-plate mold. The most common three-plate mold uses a floating plate between the A and B plate to house the runner system, as shown in Figure 3-20. The primary parting line is now between the B and X plate,

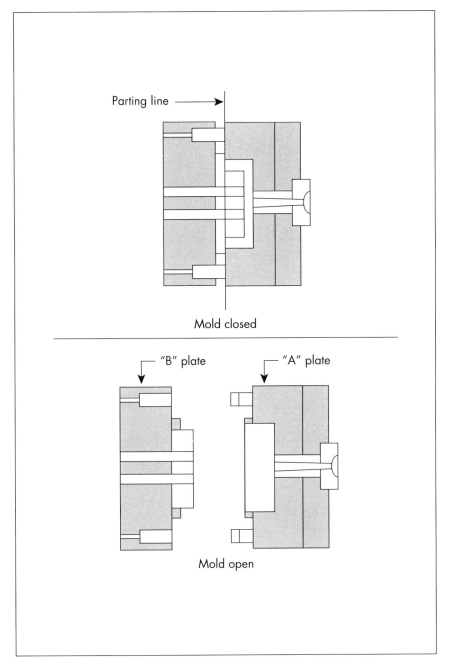

Figure 3-18. Typical primary parting line.

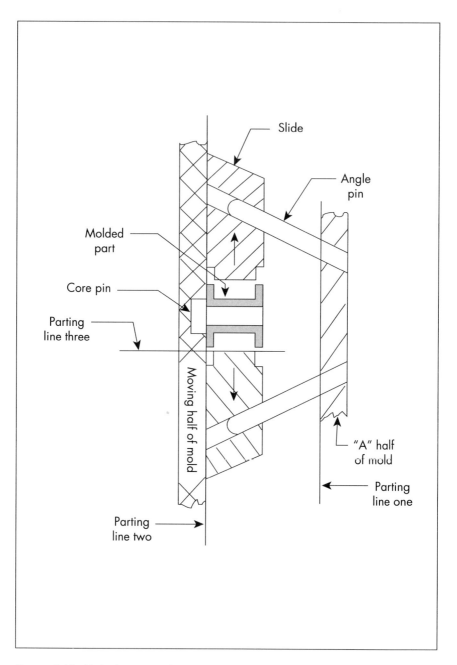

Figure 3-19. Multiple parting line situation.

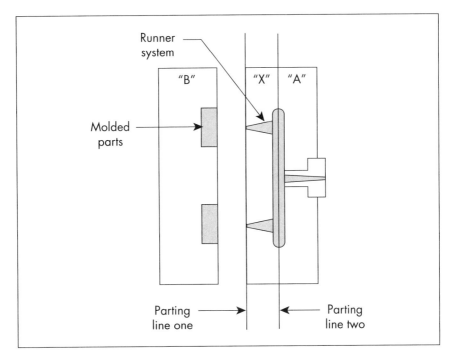

Figure 3-20. Three-plate parting lines.

and a second parting line is between the X and A plate. The secondary parting line does not actually form part of the part, but does form part of the runner system, which can be considered part of the part, by stretching our definition.

Shut-off Areas

There could be many places in a mold where components come together to form a seal while molding the finished product. These areas are called *shut-off* areas. The most common shut-off area is the primary parting line. This basically horizontal plane is commonly referred to as the "shut-off land surrounding the cavity," and is shown in Figure 3-21. Unusual contours may result in a nonhorizontal shut-off condition, as shown in Figure 3-22.

The vertical shut-off condition causes an area of concern because of the potential for damage that may occur there. This damage could result if the mating shut-off angles are not perfectly matched, or if plastic flash, or other debris, should get trapped between the angles when the mold closes. In addition, there is a tendency for the mating shut-offs to gall or seize as they slide together to seal in the last fraction of an inch during mold closing. Vertical shut-off conditions should be avoided if possible, and utilized only if absolutely necessary.

One way that vertical shut-offs can be used is to construct openings in the side walls of products without using cams or slides, as shown in Figure 3-23. Note how splitting a localized shut-off between the A and B halves of the mold will result in an opening through the wall. The shut-off area should be made with inserts that allow easy adjustment and repair in the future, if the expected production volumes warrant the additional expense.

Regardless of how the shut-off is accomplished, there should be a planned interference fit of approximately .001 in. (0.03 mm) in constructing shut-off components. In addition, all shut-off areas should be designed such that interference can be adjusted (for example, through shims) after the mold has been running for a few hours or so, due to natural "settling" of the mold components.

Machining Methods

There are many different ways to create details for the complete injection mold. Some of the components can be purchased in a finished or semifinished state from mold base suppliers, such as DME (Detroit Mold Engineering, 29111 Stephenson Highway, Madison Heights, MI 48071), or Mold Base Industries (7450 Derry Street, Harrisburg, PA 17111). But the final cavity and core sections, and details specific to a particular product design, must be made using equipment designed to produce accurate and precise components. Moldmakers have at their disposal a variety of methods and equipment to accomplish this feat.

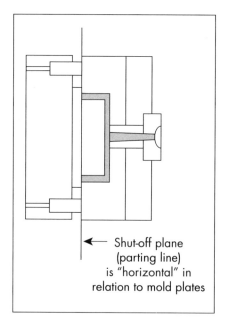

Shut-off plane
(parting line)
is "horizontal" in
relation to mold plates

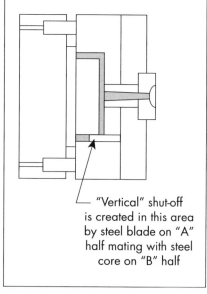

"Vertical" shut-off
is created in this area
by steel blade on "A"
half mating with steel
core on "B" half

Figure 3-21. Horizontal shut-off land.   *Figure 3-22. Vertical shut-off condition.*

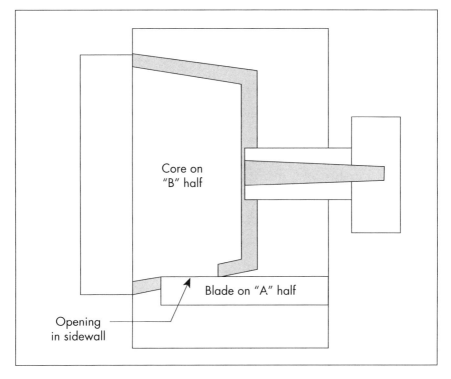

Figure 3-23. Side hole created by special shut-off.

Standard equipment includes milling machines (horizontal and vertical), surface grinders, cylindrical grinders, lathes, and drilling equipment (horizontal and vertical). If a moldmaker does not possess any of these pieces of equipment, the work can be subcontracted to a company that does. Hole drilling for waterlines is an example of this. The process usually requires a horizontal gun drill (or boring machine) for accuracy, and some machine shops specialize in providing this service to moldmakers who do not wish to tie up an investment in the special drilling equipment that might only be used once in a while.

In most cases, the cavity and core sections are fabricated using a model, either actual or virtual. An actual model is created (in wood, plaster, epoxy, or other easily machined or cast material) with the exact shape and contours of the expected product design, sometimes with shrinkage factors applied. In other cases, the shrinkage factors are determined separately during the machining phase and the model is created to the exact required finished dimensions instead. A virtual model is a computerized three-dimensional drawing. The details of the drawing are downloaded as computer files to a machining center computer that converts the files to cutting paths used by the moldmaking equipment. In this case, the shrinkage factors are usually calculated by the machining center computer. When utilizing either type of model, it is a good practice *not* to include shrinkage factors in the original model. This is because the final material selection for molding may change (or another could be added) and it will probably have a different shrinkage factor. If the shrinkage factor is developed during the machining phase instead of the design phase, the same model can be used for any plastic material.

The mold designer should be familiar with the machining methods used to produce mold components. The feeds and speeds of such equipment are detailed in many machining reference books and will not be covered here. The mold designer (and/or novice moldmaker) is prompted to refer to these information sources for a detailed analysis of how to machine mold components. However, in the following section, we will take a cursory look at some other methods of producing cavity and core components.

Other Fabrication Methods

This section will discuss fabrication methods such as casting (ceramic), hobbing, electrical discharge machining (EDM), electrochemical machining (ECM), chemical etching, electroforming (electrolytic deposition), vapor-forming nickel shells, and stereolithography apparatus (SLA).

Casting (Ceramic)

Cast tooling is usually reserved for product designs having complex shapes and contoured configurations, as well as for surfaces that are decorative (textured or grained) and difficult to attain using the more conventional fabrication methods. Moldings that must reproduce anatomical features or flora detail benefit by utilizing the duplicating accuracy of cast molds.

The ceramic casting process requires a "natural" master model, usually created in an easily machined material such as plaster, wax, soft wood, or low-melting plastic. This model must incorporate proper shrinkage factors for the plastic material to be molded. A negative pattern is then taken from this master using a cold-curing silicon rubber (or similar material). This pattern is then placed in a mold box into which liquid ceramic is poured. The ceramic compound consists primarily of zirconium sand powder mixed with a liquid thermosetting bonding agent that causes the compound to cure and harden. After a post-cure process of baking in an oven, the pattern is ready to use in a conventional metal casting process.

Hobbing

Hobbing is a method of forming molds or cavity sets in metal without removing material. A hardened and polished metal hob (master) built to the external contour of the molding, is slowly forced into a blank of soft steel at a rate of from .031 in. (0.79 mm) up to .375 in. (9.5 mm) per minute. The pattern then becomes a negative image of the hob. Preheating the blank will result in an increase in hobbing depth. Normally, the hobbing depth is limited to one times the diameter of the hob, but this changes to 1:1.5 for the cross-sectional area if the hob is not cylindrical.

As the hob is forced into the blank, a strain hardening of the blank material occurs and anneals the block, one area at a time. The surface of both the hob and the blank must be kept clean at all times or a scale will build up and interfere with hobbing. However, a lubricant should be used to allow consistent flow of the blank material during hobbing. Oil does not provide adequate pressure resistance and molybdenum disulfide is more effective. To reduce friction further, the hob is frequently copper-plated after being highly polished.

Hobbing is usually reserved for applications requiring small cavities with little height and can be more cost-efficient than casting. In addition, a single hob can be used to make several cavity sets within a very short period of time.

Electrical Discharge Machining (EDM)

The EDM process evolved from creative problem solving in the early automotive industry. A machine was developed to create an electrical spark that caused broken bolt studs to erode while still stuck in their surrounding engine blocks. This helped reduce downtime on assembly lines and the machine became known as a "sparker." The process was refined until the moldmaking industry began using it to remove the bulk of the metal during mold-building exercises. The process is shown in Figure 3-24.

A conductive material (usually carbon or graphite) is used to fabricate an electrode. This electrode is shaped to duplicate the product that will eventually be molded, and must include proper shrinkage factors and draft allowance. When the electrode is brought close to the metal of the mold cavity, a sparking condition occurs between the two objects because they are both connected to an electrical power source. The power is provided in a pulsing action that creates the spark at an average on-off frequency of approximately 40,000 cycles per second. This results in a metal removal rate ranging from .03 to .1 in.[3] (19 to 64 mm^3) per minute. The slower the removal rate, the better the final finish of the cavity set. In fact, EDM can be used to create a finish just short of that required for lenses. The average relative cost of using EDM is 75% that of fabricating the same cavity set using conventional machining methods.

Wire EDM is a method of metal removal utilizing the EDM process, but having a continuous wire (.001- to .012-in. [0.03- to 0.3-mm] diameter) that acts as the electrode. This wire feeds through a hole in the steel to be fabricated. As the wire moves through the hole, it is drawn away from the hole in the direction needed to form the shape or contour desired. In some cases, the lead hole can be produced using modern EDM equipment.

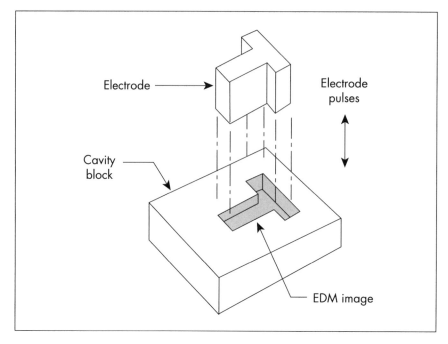

Figure 3-24. The EDM process.

Electrochemical Machining (ECM)

This process is based on using the principle of electrolytic action to dissolve the conductive material of a workpiece. The dissolving is caused by an exchange of electrical charges (and materials) between the workpiece anode and the tooling cathode. This process is rarely used for creating mold cavity sets because of the expense of producing the anodes, and the lack of accuracy available for close-tolerance parts.

Chemical Etching

The etching process has become a popular method for applying textured surfaces to mold cavity sets. These surfaces produce a finish on the molded part that ranges from a soft satin appearance to one of rough alligator hide. There are literally thousands of grades in between.

These textures are desired for a variety of reasons, including changing the appearance of an otherwise dull finish, adding a feeling of warmth to the touch of a plastic part, and hiding imperfections, such as weld lines and blush marks due to processing difficulties.

The success of the process is based on the fact that metals dissolve in acids, bases, and salt solutions. Metal atoms are forced to emit electrons and are discharged as ions from the metal lattice. These free ions are used up by reducing processes with cations and anions that are present in the etching agent. The removed metal combines with anions to form an insoluble metal salt, which is then removed from the etching agent by filtering.

The exact formulas for etching agents are well-kept secrets of the processor, and almost all mold steels and most nonferrous metals can be chemically etched. The two primary methods of etching (spray and dip) are capable of removing metal at a rate ranging from .0004 to .0035 in. (0.010 to 0.089 mm) per minute, which can be increased by heating.

It is important to understand that textured surfaces (especially on sidewalls) act as undercuts. If the texture is present on a sidewall, an additional draft allowance must be made that amounts to an extra 1° (per side) for each .001 in. (0.03 mm) of texture depth. This ensures that the molded part will be able to be ejected from the mold cavity set.

Electroforming (Electrolytic Deposition)

Cavity sets can be created by electroplating a metal shell (usually nickel) onto the surface of a conductive pattern. The thickness of the shell is in the range of only .125 to .25 in. (3.2 to 6.4 mm), but the process is slow and can take up to 4 weeks to complete this deposition.

After the shell is created, it is removed from the pattern (which can be reused) and placed on a thick backing. The backing material shape and size depends on how the finished mold will be used. For injection molding, the backing is usually a regular steel mold insert placed directly into a mold base. Aluminum also can be used if compression forces are not too high.

Electroforming is time-consuming but very inexpensive (most of the time is spent waiting for electrolytic action to be completed). Waterlines can be incorporated in the backing material if production runs are expected to be longer than a few cycles. The major advantage of electroforming is that it can be used economically to produce parts with complicated parting lines and unusual surface features (such as doll faces). Other methods of achieving these goals are too cost-prohibitive in most cases.

Vapor-forming Nickel Shells

Another method of producing nickel shells for molds is by chemical vapor deposition. While this process is over 100 years old, recent demands for inexpensive molds for the reaction injection molding (RIM) process have brought it to the foreground again.

Chemical vapor forming is a deposition process in which nickel is plated by allowing the vapors of a nickel carbonyl gas to come in contact with a heated mandrel (the plating object) that receives a uniform layer of nickel from the decomposed vapors. After a relatively short period of time (it plates at up to 0.010 in. [0.25 mm] of thickness per hour) a shell is created that accurately reproduces the finish and shape of the mandrel. The shell is then bonded to a backing similar to that used in the electroforming process mentioned earlier.

Stereolithography Apparatus (SLA)

Stereolithography apparatus utilizes the heat from an ultraviolet (UV) radiating laser to cure the surface of a pool of liquid polymer that is sensitive to UV radiation. It is similar to the process of printing on a paper surface by a laser printer. The cured portion of the plastic liquid is only .005-in. (0.13-mm) thick. This layer is then submerged .005 in. (0.13-mm) and another layer is fused to it. This process continues until a finished part is made. The finished part can then be used as a model or pattern for casting or machining a mold cavity set. The cavity set is then placed in a standard mold for the standard injection mold process.

SLA is also utilized with a special epoxy material that is used as the liquid plastic pool. A shell is formed and then cured into a very hard material. It can be back-filled with aluminum-filled epoxy and used as is for a short-run (100 to 500 pieces) cavity set, or plated with nickel to produce a high hardness for long runs (1,000 pieces or so). Most common plastic materials can be used for molding in these epoxy molds and standard gating and runner designs can be utilized. In addition, cycle times, pressures, and melt temperatures can be very close to standard injection molding parameters. This system allows molds to be built in approximately three days from the finish of a product design concept using standard computer-aided design (CAD) systems.

Miscellaneous

There are a variety of other methods employed for making cavity sets and molds for all stages of production ranging from one piece to millions annually. The reader is advised to use standard library methods to find sources for these methods, and is also advised to read the SME book, *Stereolithography and Other RP&M Technologies* by Paul F. Jacobs, and *Advanced Composite Moldmaking* by John J. Morena, published by Krieger Publishing Co., and available from the Society of Plastics Engineers, 14 Fairfield Drive, Brookfield, CT 06804-0403.

Polishing and Finishing

Usually, after a cavity set has been created through one of the many methods detailed earlier, it must be further prepared by producing a surface finish acceptable for molding the specific product. The molten plastic material used in the injection molding process will faithfully reproduce whatever finish exists on the surface of the cavity set. There are many methods used to measure the "roughness" or finish of a mold cavity, and comparative standards have been established to define this roughness. Our industry usually uses the SPI/SPE Mold Finish Comparison Standard, which has undergone recent changes. Tables III-1 and III-2 show the older version first (with equivalent expanded version values) and the expanded version second.

When specifying finishes on the mold design drawing, we usually use different finishes for the core versus the cavity. The *cavity* is usually the surface that molds the visible outside surface of the part. This would normally require a *B* series finish or higher, while the core could utilize a *C* series finish because it is not visible. Because the cost of finishing the mold surface is appreciable, we want to use only that level of finish that gives the intended (or required) result.

Polishing and finishing are both considered to be an art in the moldmaking industry. They must be performed in such a way as to result in proper "draw" to allow the finished molded product to be ejected from the mold without scratches or imperfections caused by the machining or polishing operations. And the finishing must be performed in stages going from roughest to finest. This requires a thorough knowledge of what is required, as well as a knowledge of the steel (or other material) used to make the cores and cavities. Many attempts have been made to automate the polishing processes, but human involvement has not been eliminated. It takes special knowledge, extreme patience, and a high level of coordination to perform polishing activities.

Cleanliness is critical in performing polishing and finishing operations on a molded surface. Accidental mixing of abrasives can destroy a costly mold finish. The equipment, tools, and supplies used in the various steps of polishing must be thoroughly cleaned before and after each step is started or completed. Felts, stones, sticks, and brushes should be kept separate and dedicated to a specific grade and type of abrasive. This will help minimize cross-contamination.

Textures

Textured surfaces on a molded part are required for a variety of reasons. Usually these requirements are generated by visual needs. For instance, a molded attaché case may require the appearance of hand-tooled leather. Or a tool grip may re-

Table III-1. Original Surface Finish Standards

Series	Source	Note
1	8000 grit diamond buff	Same as expanded version A-1
2	1200 grit diamond buff	Same as expanded version A-3
3	320 grit paper	Same as expanded version B-3
4	280 grit stone	Same as expanded version C-3
5	240 grit, dry blast 5 in. (13 cm) at 100 psi	Same as expanded version D-2
6	24 grit, dry blast 3 in. (8 cm) at 100 psi	Same as expanded version D-3

Courtesy of DME Company

Table III-2. Expanded Surface Finish Standards

Series	Source	Roughness	
A-1	Grade #3 diamond buff	0-1 min.(0-25 mm)	
A-2	Grade #6 diamond buff	1-2 min.(25-50 mm)	
A-3	Grade #15 diamond buff	2-3 min.(50-75 mm)	
B-1	600 grit paper	2-3 min.(50-75 mm)	
B-2	400 grit paper	4-5 min.(100-125 mm)	
B-3	320 grit paper	9-10 min.(175-250 mm)	
C-1	600 stone	10-12 min.(250-300 mm)	
C-2	400 stone	25-28 min.(625-700 mm)	
C-3	320 stone	38-42 min.(950-1050 mm)	
			Application
D-1	Dry-blast glass bead, #11	10-12 min. (250-300 mm)	8 in. (20 cm) distance at 100 psi; 5 seconds
D-2	Dry-blast #240 oxide	26-32 min. (650 mm)	5 in. (13 cm) distance at 100 psi; 6 seconds
D-3	Dry-blast #24 oxide	90-230 min. (2250-5750 mm)	6 in. (15 cm) distance at 100 psi; 5 seconds

A-series are the most time consuming and costly. They are used for mirror or optical finishes. *B-series* are used to remove all tool and machining marks. They produce good light reflecting surfaces on molded parts and provide good mold release. *C-series* are also used to remove all tool and machining marks and provide good release, but they produce a matte finish on the molded part. *D-series* are normally used for decorative parts. They produce a dull, nonreflecting finish on the molded part that helps hide imperfections, such as sink marks.

Courtesy of DME Company

quire a rough surface to minimize slipping from the operator's hand. But there are other uses for textures, such as in the case of refrigerator doors where fingerprints and dirt are hidden through the use of texturing. A popular use of texture is to hide molded imperfections, such as sink marks and knit lines.

The application of a texture to a cavity surface (there is rarely a requirement to texture a core) can be performed in many ways, including machining, EDM, and chemical- and photo-etching. The photo-etching process is the most popular. Thousands of specific textures are available and the texturing company can provide sample plaques depicting everything from a very fine matte finish to a rough alligator hide appearance. A particular finish may appear differently on a plastic material that is not the same type, color, or gloss as the sample plaque. While it is not always practical, it is wise to try to obtain a sample plaque in the exact material being considered for the final molded part.

Whatever method is used for applying the texture finish to the mold, the rougher the required appearance, the deeper the texture must be. For every .001 in. (0.03 mm) of texture depth, an additional 1° of draft must be allowed on the sidewall of the cavity to ensure release of the molded part. Most textures are in the range of .0005 to .002 in. (0.013 to 0.050 mm) in depth.

Plating

Two types of plating are commonly used in our industry: nickel and hard chrome. Nickel is used where a deterrent to wear is required, or a slight build up of the mold surface is needed. Chrome is used primarily as a thin, hard surface deterrent to wear, or to provide extra mold release characteristics to the cavity steel. There are advantages and drawbacks concerning either metal, and nickel has proven to be a more efficient and cost-effective material than hard chrome. Nickel tends to follow the surface of the mold exactly, where chrome tends to build up on sharp corners. While both materials are hard, chrome is harder, but this characteristic also makes it more brittle than nickel, and it tends to peel away easier. Chrome has a tendency to cause hydrogen embrittlement (and subsequent cracking), which is nonexistent with nickel. Chrome cannot effectively be applied at more than a .0001- to .0002-in. (0.003- to 0.005-mm) thickness, while nickel can be applied up to .001 in. (0.03 mm) or more. For replating purposes, nickel can be stripped and replated much less expensively and easier than chrome.

Regardless of which metal is used, plating is one of those arts in our industry that must be left in the hands of the experienced. Each plating company has its own specialty. Do not use a heavy plater, such as those for plating car bumpers, for mold plating.

SUMMARY

Through the application of injection and clamp pressures, mold components are exposed to tremendous forces that tend to twist, compress, warp, and bow them. These forces must be controlled by proper design and location of the components in the mold.

A core is any component having a shape that is convex (protruding) and a cavity is any shape that is concave (depressed), and together they make up a cavity set forming the cavity image.

While more expensive than cutting-in-the-solid, laminar construction of cavity sets provides ease of venting and allows closer dimensional tolerances in the finished product.

To minimize breakthrough conditions, waterlines should be located so there is steel surrounding them to at least two times their actual diameter.

A support plate is placed behind the B plate of the mold to provide supportive force against the high pressures of injection.

A single line of support pillars can allow four times the projected molding area of a mold without support pillars. Two rows can allow up to nine times the area.

A parting line can be defined as any part of the mold that parts and forms part of the part.

The shut-off area is in the form of a wall that surrounds the cavity image and provides a land over which clamp pressures are exerted.

Vertical shut-offs create a condition of concern because of the potential for damage resulting from mating shut-off contact.

There are many ways of producing cavity images. Other than basic machining, these methods include casting, hobbing, electrical discharge machining (EDM), electrochemical machining, chemical etching, electroforming, vapor forming of nickel shells, and stereolithography apparatus (SLA).

Polishing and finishing methods are employed to produce a surface finish acceptable for the specific plastic product being molded.

Textured surfaces are used to provide aesthetic appeal or comfortable feel to a molded part, as well as to hide surface imperfections, such as knit lines or blush conditions.

Because texture creates an undercut condition in sidewalls, a $1°$ per side draft allowance must be provided for each .001 in. (0.03 mm) of texture depth in these areas.

Two types of plating are commonly used in our industry: nickel and hard chrome. Nickel is used where a deterrent to wear is required, or a slight build up of the mold surface is needed, and chrome is used primarily as a thin, hard surface deterrent to wear, or to provide extra mold release characteristics to the cavity steel.

QUESTIONS

1. What action exposes mold components to forces that twist, compress, warp, and bow them?
2. What makes up a cavity set?
3. What is the term for a cavity set that is placed within a block and then attached to the very surface of the A and B plates?

4. What is the term for a set of machined cavity sets as separate blocks that are inserted into pockets and machined into the A and B plates?
5. Why are tapered wedges utilized when the pocketed cavity set method is used?
6. To prevent breakthrough, how much steel should surround waterlines?
7. How much force is it possible to generate against the B plate during the injection process?
8. If a mold base is 11.875 by 15 inches in nominal size, how many square inches of cavity can be molded if there are no support pillars?
9. Referring to question eight, how many square inches can be molded if there is a single row of support pillars added?
10. How many pieces does the ejector housing consist of and what is its shape?
11. What do the initials EDM stand for?
12. What do the initials SLA stand for?

Action Areas of the Mold   4

OVERVIEW

Most molds are not just blocks of metal with a cavity image. They usually have some type of mechanism for the forming of the molded product during the injection process, or its release from the mold after the process is completed. These mechanisms, called *actions,* must move sometime during or after the molding process, and include devices such as slides and cams, unscrewing mechanisms, and ejector systems. We will look at some common ones as well as hand-loaded inserts.

SLIDES

A product design with features that are parallel to the primary parting line of the mold (but below it) may require an action device in the mold. Some examples are: an undercut resulting from a thread requirement, a "zero draft" condition on a sidewall, or a large window-style opening in the side of a box-shaped part. They all require a method for moving a section of the mold out of the way to eject the product. If an entire section must move, we refer to that section as a *slide,* because it is said to "slide" out of the way. If only a small portion of the section must move, we perform that activity using a cam, which is detailed following the discussion on slides.

A typical slide mechanism is shown in Figure 4-1. Note that the face of the slide is an integral portion of the cavity image and actually forms an entire section of the molded part. When the slide is in the forward position, it shuts off against the rest of the cavity set and creates a sealing area to contain the incoming molten plastic. When the slide is in the back position, the face is moved out of the way so it will not interfere with the ejection of the molded part from the mold. The slide can be activated by various means including angled pins, hydraulic cylinders, and rack-and-pinion devices. By far the most common method is by using angled pins. These are also called *horn pins* because they have a shape similar to animal horns, as shown in Figure 4-2.

The pin is stationary in the mold and placed at an angle (no more than 25° from vertical) to the parting line. The slide has an opening in it that accepts entry of the angled pin. As the mold closes, the slide is forced to move forward as it follows the path generated by the angled pin. The angle and length of the pin are

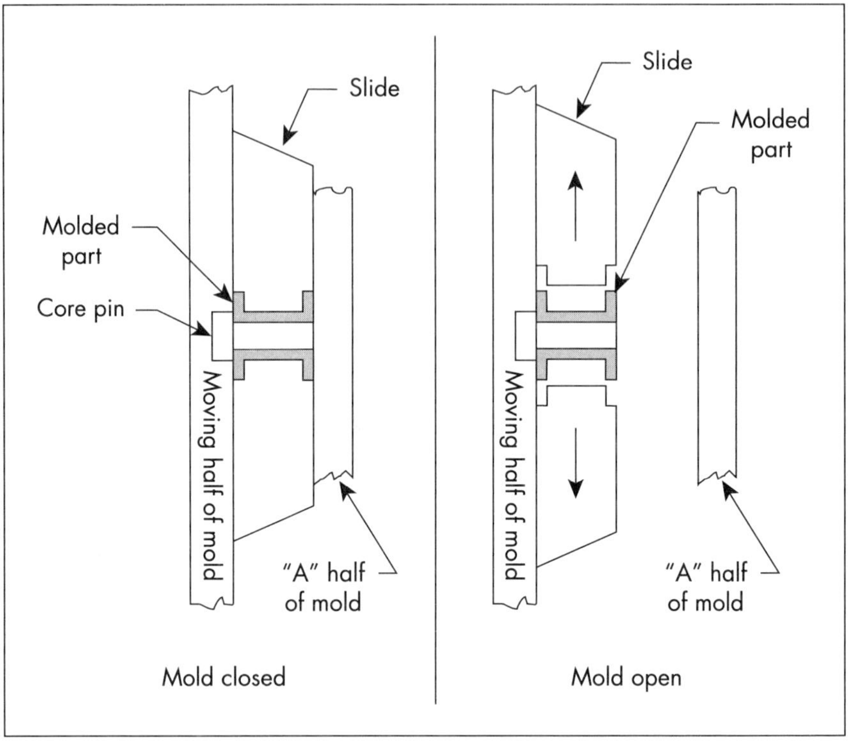

Figure 4-1. Typical slide mechanism.

determined (timed) to ensure that the slide travels until it is tight against the shut-off area formed between its face and the rest of the cavity set. Final timing can be achieved by wear plates (for metal buildup/removal if necessary) between the face of the slide and the block to which the face is attached.

The mold should not be allowed to open to the extent that the angled pin ever leaves the confines of the hole in the slide. If that were to happen, the slide may accidentally move and, upon the mold closing, the pin would not enter the hole, but contact the upper face of the slide and cause a great deal of damage. In fact, safety devices are utilized to minimize this possibility. First, a stop is added to the area at the rear of the slide travel to ensure the slide cannot move beyond the intended rearward travel. Then, a spring-loaded plunger device ("Vlier" button) can be placed against the bottom of the slide. This ball-shaped plunger will slip into a small cavity machined into the bottom face of the slide exactly at the point at which the slide should stop when the mold is fully open. An electrical contact switch can be added that must be activated by the slide before the molding ma-chine is allowed to close the mold. These are typical safety methods and there are

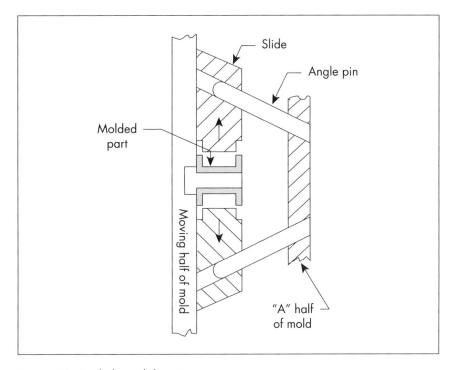

Figure 4-2. Angled pin slide activator.

many more that can be used. The idea is to keep the slide in the proper position at all times during the molding process.

CAMS

Cams, as shown in Figure 4-3, are similar devices to slides because they are used for creating features in the sidewalls of a molded part parallel to the primary parting line. However, cams normally do not travel very far and usually form only a small feature (such as a round hole) in the side of a molded part. Thus, the cam might contain only a core pin that would travel in and out of the sidewall of the cavity image. This would create a hole in the sidewall of the part. Of course, the feature could be something other than a round hole, but the basic concept of a cam is that it provides only a single feature item.

The actuation method of the cam varies with specific mold designs. The two most common methods of actuation are a tapered block and a small angled pin. These are shown in Figure 4-4.

Usually there are locating devices such as Vlier-style plungers and/or springs to keep the cam properly positioned at the varying stages of activation.

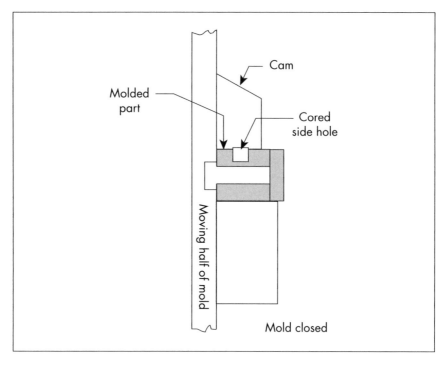

Figure 4-3. Typical cam mechanism.

LIFTERS

The cam can form a feature from the outside of the part or from the inside of the part. What we have discussed so far is external cam action. If a cam is used for forming a feature on the internal portion of the molded part, it must be activated by a system similar to what we call a *lifter,* as shown in Figure 4-5. This device is attached to the ejector system so that it not only moves back and forth to form the internal feature, but also travels up with the part as the ejector system pushes the part out of the mold and back as the ejector system is positioned for the next cycle. Lifters are valuable devices for creating otherwise impossible-to-form internal sidewall features. The vertical included angle of activation, as with standard slides, should not exceed 25° to keep wear at a minimum level.

UNSCREWING DEVICES

Sometimes, undercut features take the shape of screw threads. In these situations, the area of the mold forming the threads must be rotated away from the plastic after being formed. The mold can be designed so that either the mold component

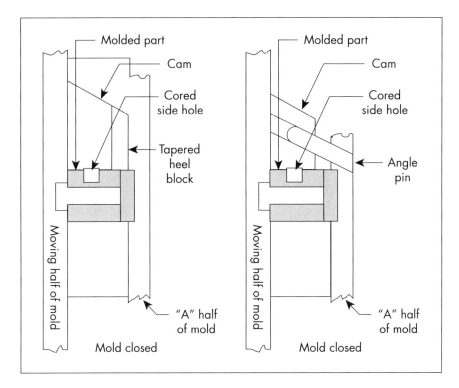

Figure 4-4. Basic cam activators.

is threaded away from the plastic part, or the plastic part is threaded away from the mold material. The most common method is to thread the mold material away from the plastic part.

First, let's look at the type of undercut condition we are discussing. Figure 4-6 shows an external and internal thread on molded plastic parts. If you consider the cap and neck opening of a plastic soft drink bottle, you will see an external thread on the bottle and an internal thread on the cap (commonly referred to as male and female undercuts). To successfully mold either of these undercuts, the metal in the mold must be moved out of the way after the plastic solidifies. This enables the molded part to be removed from the mold.

Unscrewing devices can be built to rotate the mold component away from the solid plastic or rotate the plastic away from the mold component. Figures 4-7 and 4-8 show common versions of these methods, although there are many ingenious techniques in use throughout our industry. Regardless of which method is utilized, the concept is to eliminate the constraining feature of the undercut condition to allow the plastic part to be ejected from the mold.

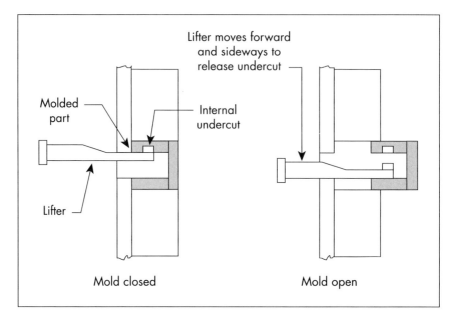

Figure 4-5. Typical lifter mechanism.

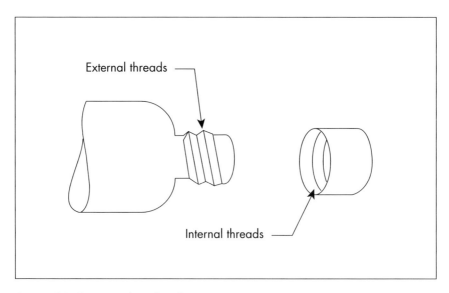

Figure 4-6. Common thread undercuts.

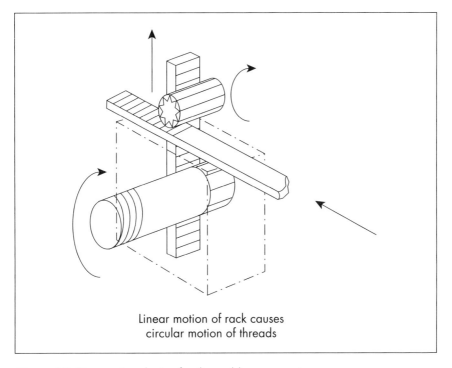

Linear motion of rack causes
circular motion of threads

Figure 4-7. Unscrewing device for the mold component.

HAND-LOADED INSERTS

Another popular method of forming threaded areas, undercuts, or other sidewall features on a molded part, is to use inserts that are placed in the mold at the beginning of each cycle and removed at the completion of each cycle. These can be placed and removed using robots or other sophisticated mechanisms, but the most common way is to insert and remove them by hand. The components that are used are called *hand-loaded inserts* and they are located strategically in the cavity set of the mold before it closes. After the mold closes, plastic enters the cavity and molds around the hand-loaded insert. The mold opens, and the entire part (including the insert) is ejected from the mold. At that time, the operator removes the hand-loaded insert from the finished part and places it (or another prepared insert) in the mold for the next cycle. While this is a labor-intensive and time consuming process, it is much more cost-effective than using mechanical unscrewing methods for molding low-volume quantities (less than 25,000 units annually).

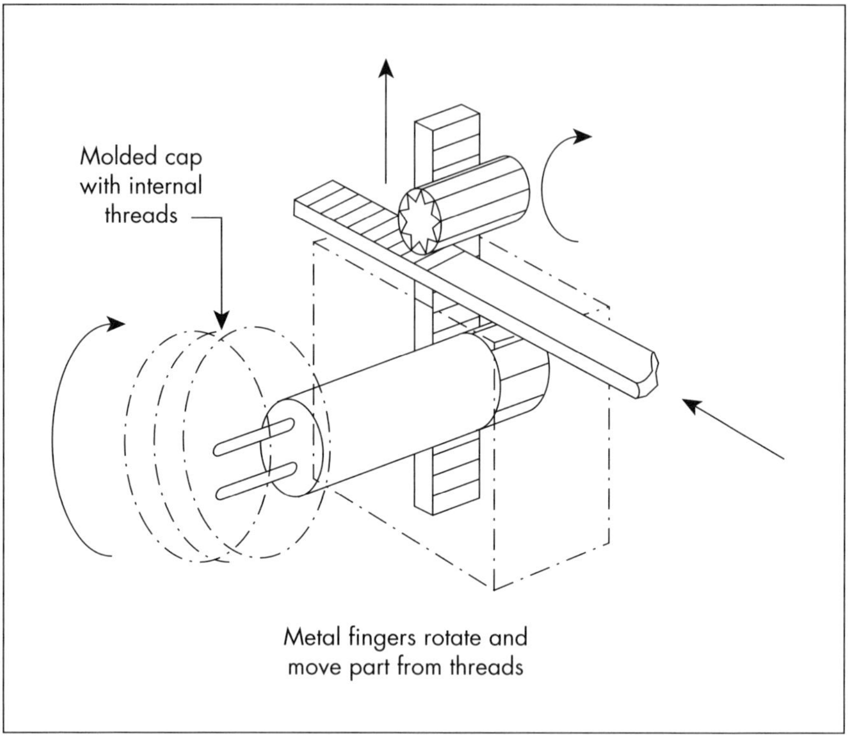

Figure 4-8. Unscrewing device for the plastic part.

EJECTION SYSTEMS

After the plastic part is molded, it must be removed from the mold. This is accomplished by using methods that push the solidified plastic from the cavity set. While there are many ways of doing this, we will look at some of the more common methods employed. This section will discuss the design of a standard ejection process, stripper ejector systems, the split cavity concept, a three-plate system with delayed ejection, and compressed air ejection.

Standard Ejection Design

In the standard design the molded part is ejected using a set of ejector "pins." The most common shape is like that shown in Figure 4-9. The pin construction consists of a head, a body, and a face, usually forged in H-13 tool steel, hardened, annealed, and nitrided. The face is normally flat but can be machined to any shape desired. These pins come standard in a variety of body diameters from 1/8 to 1 in. (3.2 to 25.4 mm), and in overall lengths from 6 to 36 in. (15 to 91 cm). They are machined to the required length and fit to a specific mold location by the

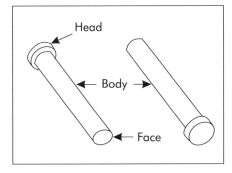

Figure 4-9. Common ejector pins.

moldmaker. To keep the ejector pins from penetrating the plastic during ejection, the ejector pin diameters should be as large as possible so that the ejection pressure is distributed over a large area. This also will help minimize distortion of the plastic part during ejection.

Ejector pins are mounted in a counterbored ejector plate (called the *retainer plate*) that has a backup plate (called the *ejector bar*) bolted behind it, as shown in Figure 4-10. This en-

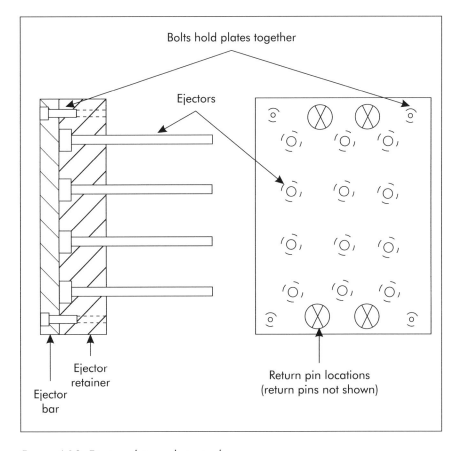

Figure 4-10. Ejector plate and ejector bar.

tire assembly is then located within an ejector housing, mounted directly behind the B plate assembly of the mold, as shown in Figure 4-11.

In most cases, the entire ejector system is guided using pins and bushings to minimize wear and distortion during use. This is especially needed for large molds or for long-running, high-volume situations. However, this concept does not need to be used for lower volume production or on small molds.

Note that in Figure 4-11 the ejector system has other pins in place. These are the return pins and sprue puller. The return pins bring the ejector system back to its home position in preparation for the upcoming cycle. This is done when the B half of the mold closes against the A half of the mold. The sprue puller pin is used to help extract the sprue from the sprue bushing when the mold first opens. Then, it is used as an ejector pin to push the sprue from the B half of the mold during the ejection process.

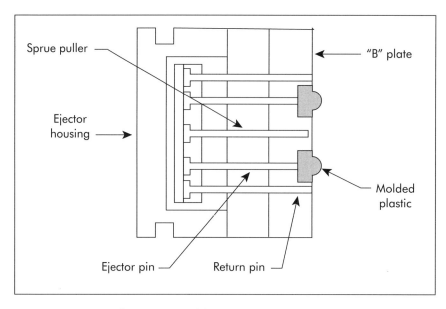

Figure 4-11. Ejector housing assembly.

The ejector pin is machined so that it is "flush" at the bottom of the cavity. However, due to tolerances required, dimensional stack-ups, and expansion/contraction of the mold plates during the molding process, it is impossible to maintain the original flush condition of these pins. Therefore, determine whether to make the pins longer or shorter to compensate for these factors. The effect will be one of two things: either the part will fall freely from the mold or the part will stick on the ejector pins and need to be removed manually (or by use of a robot device). The reason for this situation will be better understood by examining Figure 4-12.

If the ejector pin is too short, it forms a pad of excess material because the incoming plastic fills in the voided area in front of the ejector pin face. Now, when the ejector pin moves forward, it pushes against the pad, and the molded part ejects from the cavity and falls freely from the mold. However, the pad of excess plastic is permanent and may be objectionable to the product designer or end user.

When the ejector pin is long, it forms a depression in the molded part because the incoming plastic material molds around the extended body of the pin. As the plastic cools, it shrinks to the ejector pin. Then, when the pin is pushed forward by the ejector system, the molded part is pushed from the cavity, but stays put on the ejector pin and does not fall from the mold.

Ejector pins must be designed to be either longer or shorter than the mold surface. If the depressions or pads from the ejector pins are objectionable, another method of ejection might need to be considered.

Stripper Ejector Systems

There are situations where standard ejector pin systems cannot be utilized either because the pins are so small that they can puncture the molded part, or there is

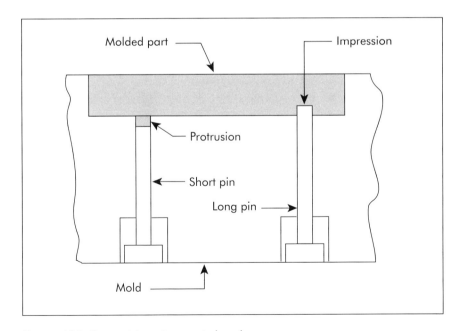

Figure 4-12. Determining ejector pin length.

limited (or no) area in which to place standard ejector pins. At times like these, we can look to stripper ejector systems to solve the problem.

There are two versions of stripper systems: sleeves and plates. These are shown in Figures 4-13 and 4-14.

A *stripper plate* is a third plate added to the main mold between the A and B plates. It is attached so that it is integral to the B half of the mold and forms the lower areas of the plastic part by replacing that section of the mold core set. As shown in Figure 4-13, when the mold opens, the part stays on the B half because it has shrunk to the core. When the ejector system activates, the return pins push the stripper plate forward. This action strips the molded part from the core and ejects the finished product from the mold. The stripper plates are available off-the-shelf as standard mold plate items from mold base suppliers.

When a stripper sleeve is used, the sleeve is actually a hollow ejector pin. The sleeve fits around a core, as shown in Figure 4-14. Again, the molded part shrinks around the core and stays on the B half when the mold opens. When the ejector system is activated, the sleeve ejector strips the molded part from the core and ejects the finished product from the mold. Stripper sleeves are available off-the-shelf from mold base suppliers.

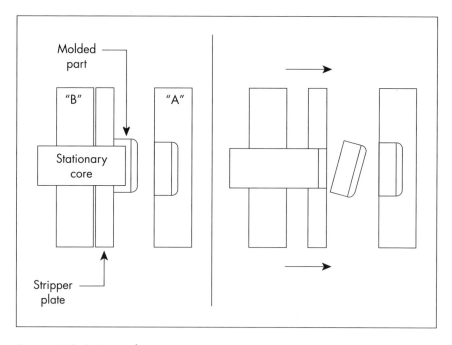

Figure 4-13. Stripper plate ejection.

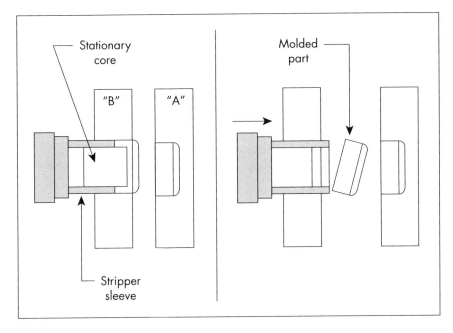

Figure 4-14 Stripper sleeve ejection

Split Cavity Concept

The split cavity concept is commonly used when molding a deep part with or without external undercuts or sidewall openings, as shown in Figure 4-15. When the mold opens, the molded part is captured in the B half. As the ejector system activates, an ejector pushes forward and causes the angle-walled cavity sections to separate. As they do, the molded part is allowed to be removed because the constraining walls have been separated. With proper draft allowance, the part will fall from the mold due to gravity. However, if there is not proper draft, the part must be forced from on whichever side it has stuck. This can be accomplished by using side ejector pins or prying from the side with special hand tools.

Three-plate System with Delayed Ejection

A three-plate mold design is used for many reasons. One is to allow automatic degating of the part from the runner system during the ejection phase of the molding cycle. Please refer to Figure 4-16 as we explain.

The third plate is placed between the A and B plates so that the runner is located between the A plate and the third plate. Drop gates are utilized to feed molten plastic through the third plate to the cavity image. Sucker pins hold the

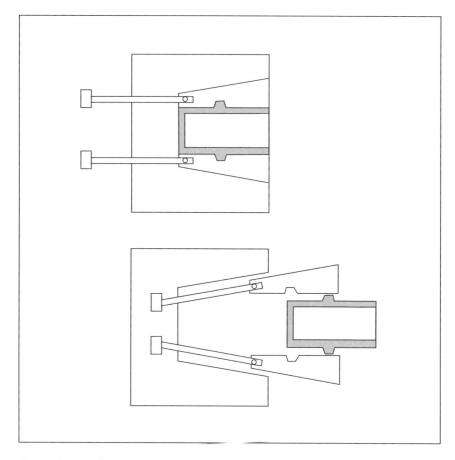

Figure 4-15. Split cavity ejection system.

runner to the A side when the mold opens, and the molded parts are forced to stay on the B side through product design or the use of small scratch undercuts. The third plate is connected to the B half of the mold by a set of slotted bars. When the mold approaches the final opening, the bars pull the third plate toward the B half of the mold. This causes the runner system to break loose from its entrapment and fall freely from the mold. Now, the ejection system of the machine activates and pushes the finished product from the B half of the mold in standard ejection fashion. The use of a baffle system (cardboard panel), placed below the mold and in the general parting line plane, can be effective in keeping the runner system separate from the molded parts as they all fall into individual boxes or onto separate conveyor systems for disposal. This system is commonly used when processing requirements call for a totally automatic molding operation.

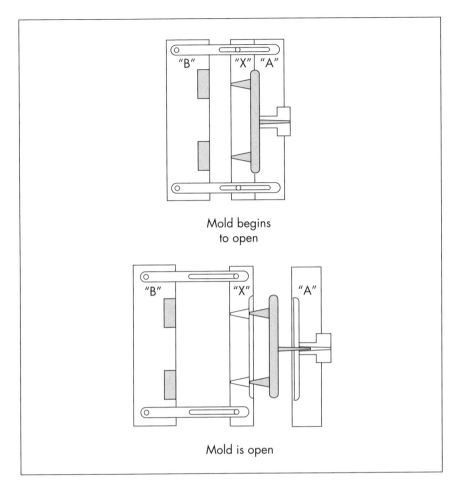

Mold begins
to open

Mold is open

Figure 4-16. Three-plate system with delayed ejection.

Compressed-air Ejection

Compressed air can be used in conjunction with a standard ejection system to remove deep-draw parts or parts that by design are difficult to push. It forces the plastic away from the core and allows it to fall freely from the mold. Figure 4-17 shows a typical example of this method.

In Figure 4-17, we show a deep, cup-shaped product after the mold has opened. At this stage, poppet valves open on the core of the mold and allow compressed air to enter between the molded plastic and the core. This forces the plastic away from the core to which it has shrunk and allows standard ejection to push the part far enough forward to make it fall freely from the mold. This process requires

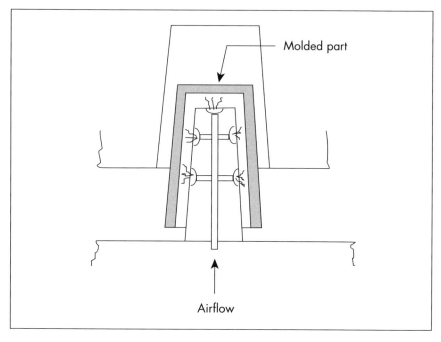

Figure 4-17. Compressed-air ejection.

exact timing between the ejection system and the compressed-air release, but is very effective in ejecting otherwise difficult-to-remove products.

SUMMARY

Most molds incorporate some mechanisms, called actions, needed for the forming of the molded product during the injection process, or the release of it from the mold after the process is completed.

If an entire section of the mold must move out of the way to eject the part, we call that section a slide as a way of defining its action. If only a small section of the subject area must be moved, we use a device called a cam.

The most common method used for activating a slide or cam is the angled pin (or horned pin) concept.

While slides and cams form undercuts on the external surface of the molded part, lifters are used for forming undercuts on the internal surface.

Unscrewing devices are commonly used for the forming of undercuts similar to screw threads, either internal or external. These devices cause either the molded part or the mold component to rotate away from the mold.

Hand-loaded inserts are the most common devices used for forming undercuts in low-volume production situations.

While ejector pins are the most commonly used devices for removing a molded part from the mold, stripper plates and sleeves may be required if ejector pins cannot be used.

Air ejector systems can be utilized for removing parts that are difficult to eject.

QUESTIONS

1. What is the general term used to define any mechanism of the mold that must move out of the way to allow molded parts to be ejected from the mold?
2. An angled pin is the most common method used for activating a slide. What is another common term used for describing the angled pin?
3. Why should an angled pin not be allowed to leave the confines of the mating hole found in the slide or cam?
4. What devices are commonly used for undercuts that are on the external surface of a molded part?
5. What device is usually used for forming undercuts on the internal surface of a molded part?
6. What is the proper angle to incorporate when utilizing slides, cams, or lifters?
7. What are the names given to the three items that make up an ejector pin?
8. Why should ejector pins intentionally be made too long or too short?
9. Which design will allow a molded part to fall freely upon ejection: too long a pin or too short a pin?
10. What device can be utilized to remove a part from a mold if ejector pins cannot be used?

Runners, Gates, and Venting 5

OVERVIEW

The injection molding process pushes molten plastic into a closed mold. The trapped air in the closed mold must first be removed. Then, the plastic must flow through the mold and enter the cavity image where it solidifies before final ejection. There are three main items that are needed for this to take place properly: a runner system, a gating system, and proper venting. In the following section, we will discuss each of these items, look at some of the variations available, and understand the requirements of each. First, let us look at the sprue bushing.

ROLE OF THE SPRUE BUSHING

The sprue bushing is part of the mold and acts as the interface between the injection molding machine's cylinder nozzle and the runner system of the mold. The plastic that is formed within the sprue bushing during the molding process is called the *sprue*. There are certain rules applied to the design of each sprue bushing, because the bushing is very specific to a particular mold design and product design.

Figure 5-1 shows how to dimension a sprue bushing properly. Note that the bushing has a spherical radius on the face of the large head end. This radius is designed to form a seal against the radius of the nozzle on the molding machine, and must be the same as that radius, or slightly larger. If the sprue bushing radius is smaller, flash will form. The sprue will stick in the bushing every cycle rather than pulling free as it is intended (some mold designs, notably for insert molding, do not require this radius, and are in fact flat). In addition, the smaller diameter of the inside tapered hole must be equal to or slightly larger than the mating hole of the machine nozzle, for the same reasons. The size of that smaller hole should be kept to a minimum, but large enough to meet the requirements of the specific plastic being molded. The nozzle diameter can be changed to suit the diameter of the sprue bushing, if necessary. The common diameters for this interface are 3/32, 5/32, 7/32, and 9/32 in. (approximately 2, 4, 5.5, and 7 mm), although *any* diameter is possible.

The inside diameter at the large end of the tapered hole is determined by adding up the cross sectional areas of all the runners leading from the sprue bushing. Figure 5-1 shows two runners leading from the sprue bushing, and each has a cross sectional area of .01227 in.^{2} (7.9 mm^2). The total for both then is .02454 in.^{2} (15.8 mm^2). Now, we must find a single circle size that would encompass at least

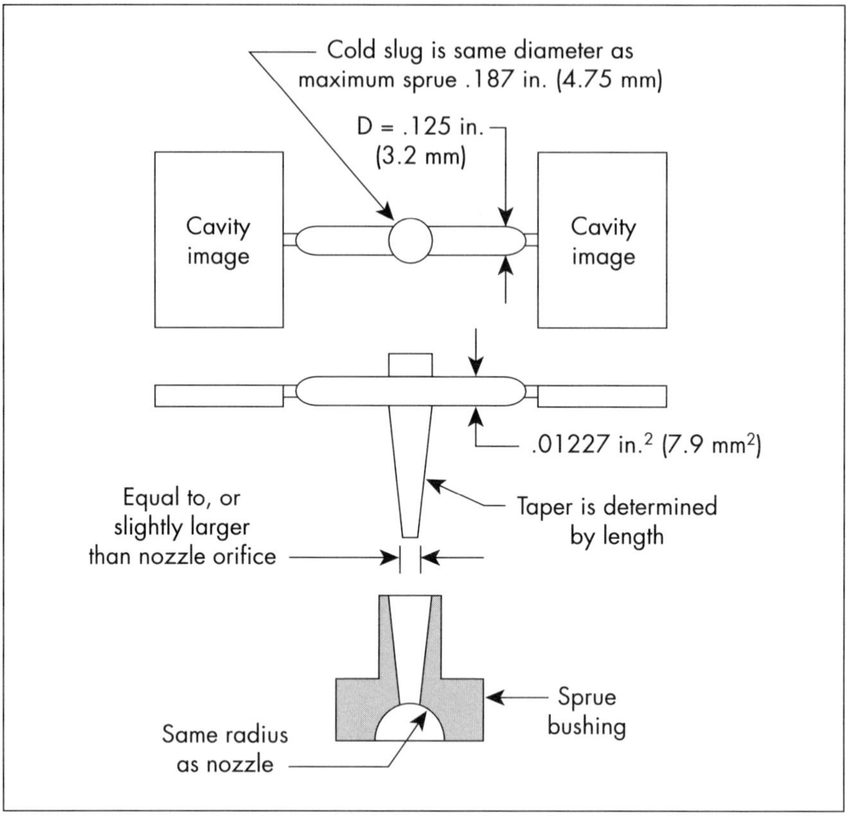

Figure 5-1. Sprue bushing dimensions.

that total (refer to the *Machinery's Handbook* [see Bibliography] for charts). The closest common circle with at least that area has a total area of .0276 in. (0.7 mm) and a diameter of 3/16 in. (4.7 mm). This would be the minimum size of our diameter at the large end of the internal sprue bushing hole.

The overall length required determines the taper of the inside sprue-bushing hole and should be kept at a minimum. Taper is measured from the outside face of the A half of the mold, through the mold plates on the A half, and ends at the intersection where the bushing meets the runner system. If this dimension is outside a common depth, an extended nozzle can be put on the machine and a shorter bushing can be incorporated.

The cold slug portion of the sprue is usually formed in the B half of the mold by machining a small hole in the B plates directly where the sprue bushing intersects the runner. This captures the first (and coldest) portion of the incoming plastic and keeps it from plugging up the flow path in the runner system or gate areas.

RUNNER SYSTEMS

Runner systems direct the molten plastic, after it enters the mold, from the sprue bushing to the cavity images that form the molded product. They are called *runners* because the plastic "runs" along the channels that are machined into the mold base. In most cases, runners are used only for multiple-cavity molds. Theoretically, a single-cavity mold does not need a runner because the plastic is injected directly into the center of the single cavity, an ideal situation. However, some single-cavity molds may be positioned in an off-center location in the mold and a runner may be required to carry material to that location from the sprue bushing. This is especially true in "family" molds, or "universal" molds where only one (or a few) cavity may be running at any one time.

Surface Runners

Surface runners are usually cut into a block called a *runner block* inserted into the A and B halves of the mold, as shown in Figure 5-2. This is done to allow adjustments and changes to the shut-off and dimensional functions of the basic runner system, without affecting the entire mold base or cavity layout.

Proper design of the surface runner dictates that a full-round runner is ideal. This is so because a circular cross section creates equal pressure in all directions on the plastic molecules, where a noncircular section causes unequal pressure. This is shown in Figure 5-3, where a full round runner design is compared to a common trapezoid runner design. Using the runner design on the left side of Figure 5-3 will minimize the amount of molecular distortion created while the molten plastic is flowing through the runner toward the cavity. The trapezoidal cross section shown on the right causes molecular distortion, which results in stresses that set up in the material. These stressed molecules are carried into the cavity where they solidify in their stressed state.

The proper runner cross section diameter depends upon the type of plastic being molded. High-viscosity (stiff) materials require larger diameters than low-viscosity plastics. Table V-1 shows normal runner diameters for some common materials. As shown, the longer the flow path the plastic must travel along, the larger the runner diameter must be at the start. For example, a polycarbonate material requires the runner diameter to start at .125 in. (3.1 mm) when the runner length is only 3 in. (76.2 mm), but requires the runner diameter to start at .156 in. (3.9 mm) when the runner length is 6 in. (152.4 mm) (as shown in Figure 5-4). This length is measured from the cavity being filled, back to the center of the sprue bushing (or center) of the mold. The reason for the required increase in diameter is that the plastic begins to cool down upon entering the runner system and will start to solidify as it moves toward the gate. Therefore, we must increase the size of the runner to allow the molten plastic to move faster along the path before it solidifies. However, the larger the runner diameter, the longer it will take to cool down and be able to be ejected from the mold. Therefore, we must carefully

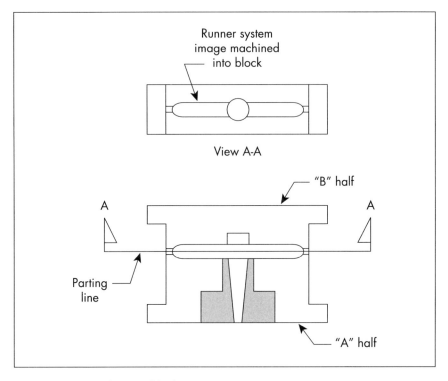

Figure 5-2. Typical runner block concept.

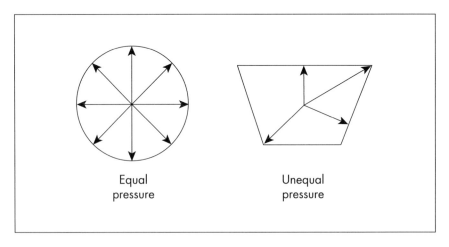

Figure 5-3. Comparing round runner cross section with trapezoid.

Table V-1. Runner Diameters for Common Materials

Material	Running Length		
	3 in. (76.2 mm)	6 in. (152.4 mm)	10 in. (254 mm)
	Runner Diameter		
	in. (mm)	in. (mm)	in. (mm)
ABS	.093 (2.4)	.109 (2.8)	.156 (3.9)
Acetal	.062 (1.6)	.093 (2.4)	.125 (3.1)
Acrylic	.125 (3.1)	.156 (3.9)	.187 (4.7)
Cellulose acetate	.093 (2.4)	.109 (2.8)	.156 (3.9)
Cellulose acetate butyrate	.093 (2.4)	.109 (2.8)	.125 (3.1)
Ionomer	.062 (1.6)	.093 (2.4)	.125 (3.1)
Nylon 6/6	.062 (1.6)	.078 (1.9)	.093 (2.4)
Polycarbonate	.125 (3.1)	.156 (3.9)	.203 (5.1)
Polyethylene	.062 (1.6)	.093 (2.4)	.125 (3.1)
Polypropylene	.062 (1.6)	.093 (2.4)	.125 (3.1)
Polyphenylene oxide	.125 (3.1)	.156 (3.9)	.203 (5.1)
Polyphenylene sulfide	.125 (3.1)	.156 (3.9)	.203 (5.1)
Polysulfone	.156 (3.9)	.187 (4.7)	.218 (5.5)
Polystyrene	.093 (2.4)	.109 (2.8)	.125 (3.1)
Rigid PVC	.125 (3.1)	.187 (4.7)	.250 (6.3)

consider the larger diameter requirement when determining the overall cycle time for molding the product. The usual case is that the runner diameter is larger than the thickest section of the part to be molded. This results in an overly long cycle time, from allowing the runner enough time to cool down sufficiently. While it does not have to cool to the state of being completely solid, it does have to cool long enough to be somewhat rigid and ejected properly. In addition, every time the runner must make a right-angled turn, the starting diameter must be increased by a factor of 20%. This is to make up for the drop in pressure when the material is forced along the runner path, as shown in Figure 5-5.

The ideal runner path takes material on a straight line from the sprue bushing to the cavity image. However, due to waterline interference, bolthole locations, and ejector pin layouts, a straight line may not be possible. In those cases, a runner is usually designed to take right-angled turns as shown in Figure 5-5.

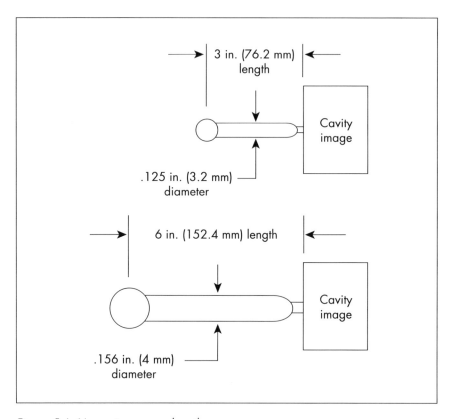

Figure 5-4. Measuring runner length.

A better way, if possible, is to use a sweeping radius design as shown in Figure 5-6. This approach eliminates the pressure drop caused by sharp turns and the need to increase the runner diameter by 20% for each of those turns. The runner diameter is still determined by measuring its length between the cavity image and the sprue, but there are no other factors to apply. This minimizes the runner size and helps lower the overall molding cycle. Steps should be taken in the early mold design stages to incorporate this concept if possible.

When laying out the cavity images of a multicavity mold in the first place, thought should be given to creating a "wagon-wheel" design, with runners taking the place of the wagon-wheel spokes, as shown in Figure 5-7. This is the best design because it uses the straight-line runner approach and minimizes the travel the plastic flow fronts must make to get to the cavity images. It also keeps the runner diameter at a minimum, thus reducing overall cycle times.

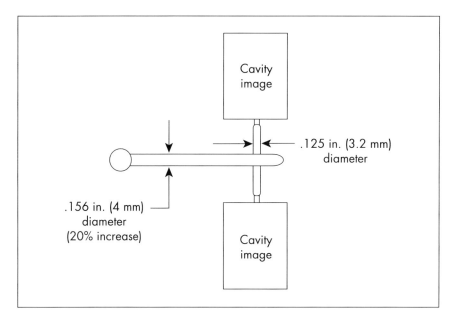

Figure 5-5. Increasing runner diameter by 20%.

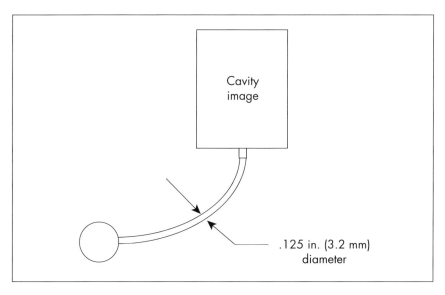

Figure 5-6. Sweeping radius runner path.

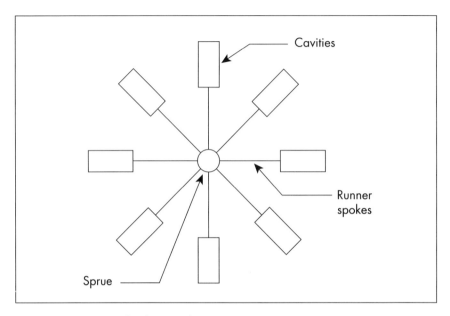

Figure 5-7. Wagon-wheel runner layout.

Insulated Runners

The long cycle times associated with standard surface runners (due to their thickness) caused molders to try to find ways to overcome the condition. One such effort resulted in what is known as an *insulated runner*. In this system, the runner is not machined into the surfaces of the A and B plates, but rather between the A plate and an in-between plate known as the X plate, or third plate.

Figure 5-8 shows the insulated runner system concept. The A plate and X plates are bolted together. The runner is machined between those two plates, and drop gates are used to bring the molten plastic to the cavity image that is located between the B and X plates. When the mold opens, the runner is trapped between the A and X plates. The molded part is ejected and the mold closes for the next cycle. The runner thickness is large and the runner begins to solidify from the outside in, but the center core of the runner stays hot and molten. When the next cycle starts, the plastic is injected through the center core of the runner and the still-molten material fills the mold. The result is that the center core area has been insulated from the cooling effect of the mold steel because of the very thick skin forming on the outside of the runner diameter. This process can continue as long as the center core of the runner stays hot enough to keep the plastic molten. A slight interruption of the cycle may be enough to cause the center core to cool and solidify. When this happens, the mold must be dismantled and the solidified

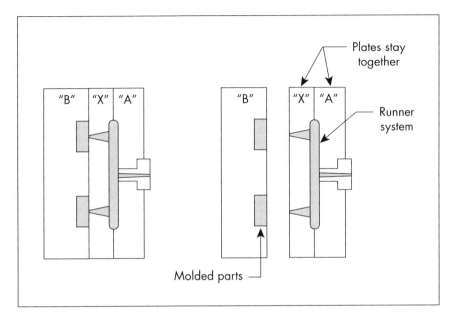

Figure 5-8. Basic insulated runner concept.

plastic must be machined out of the runner system. Then, the runner must be polished and the mold can be reassembled for another molding attempt.

The major advantage to the insulated runner system is that the runner does not have to be included in the calculation of cycle times. The cooling portion of the molding cycle only applies to the molded part and the overall cycle can be much shorter than if runner diameter thickness were included. A secondary advantage is that there is no runner to dispose of after molding. That eliminates the need to use or inventory *regrind* (recycled process resin) that would normally be generated by a standard runner system.

Hot Runners

The inefficiencies and uncertainties associated with insulated runner systems led the way to creation of a hot runner system. The purposes for both systems are the same: eliminate the surface runner system and reduce overall cycle times. However, the hot runner system uses individual cartridge heaters to keep the plastic molten in the runner path and does not rely on the insulating properties of the runner skin. In fact, as Figure 5-9 shows, the runner skin never does solidify. However, the entire hot runner system is insulated from the rest of the mold to keep the basic mold temperature comparatively low while the runner temperature can be high.

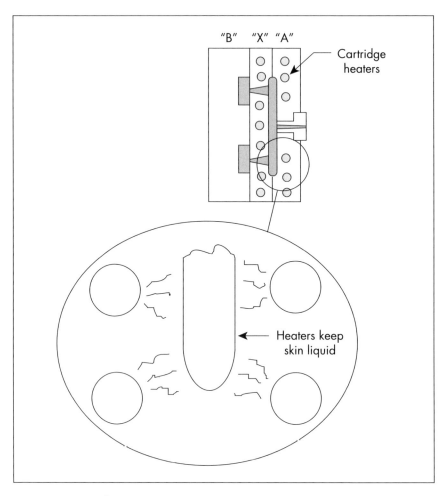

Figure 5-9. Basic hot runner concept.

The hot runner concept has allowed a remarkable thing to happen. In effect, the nozzle of the molding machine has been moved directly to the cavity image, thereby eliminating the stress and control conditions normally found in the use of a standard surface runner system. This, in turn, results in some major advantages. Cycle times are shorter by an average of 25% because there is no runner to include in the calculation, defects are minimized as a result of minimizing stresses, and scrap is reduced because there is no runner to dispose of after molding.

Hot runner systems are commercially available, but must be engineered for each mold and material on a specific basis. They are expensive, with the average cost in the area of $25,000, but they pay for themselves quickly due to faster cycle times and less scrap. Even heat-sensitive materials, such as polyvinyl chloride (PVC), can be run in modern hot runner systems. If the investment can be justified, hot runner systems should be considered for every mold.

GATING METHODS AND DESIGNS

After molten material travels through the runner path, there are many ways to get it into the cavity image. The runner itself can continue directly into the cavity image, and, in fact, that was the method used in the very early days of our industry. However, that made it difficult to remove the molded part from the runner system, due to the mass of material at the junction formed by the runner and part. Therefore, a reduction in runner size was created at this junction. This reduction became known as the *gate* through which material must travel to enter the cavity image. The reduced area allowed easier removal of the molded part from the runner. In addition, this reduced area created a slight amount of friction that caused the plastic going through it to heat up. This extended the flow of the plastic material and made it easier to fill the cavity image. Today, the gate is used for two purposes: to control the flow of molten material entering the cavity image and to ease separation of the molded part from the runner system.

Early gate designs were primitive and not scientific in nature. Most gate sizes were estimated and alterations were made after running the mold to see the results. However, as more materials became available for molding (some of which were heat-sensitive), and as molding machines became more controllable, gate design became an important issue. It was determined that specific materials could have wide molding parameters, while others needed very tight control, especially in the area of gate design. A gate that was too "tight" might cause thermal degradation and stress to be molded into a part. However, a gate that was too large might result in excessive cycle times and difficulty in removing molded parts from runners. Gate design, then, became a science in itself and resulted in a variety of shapes and concepts, some of which we will investigate in the following section.

Determining Gate Location and Number

First, determine where the part should be gated and how many gates might be needed. A statement can be made that says, "any part can be filled with a single gate." While this is true, it might be better to add gates to overcome some of the problems associated with just a single gate, depending on product design and final requirements of the molded part. For instance, if absolute flatness is required of a molded part that has varying wall thickness, a single gate may not be able to produce such a part due to internal distortions caused by the flow of material. Other factors may also affect the number and location of the gates.

Most parts are not designed to have uniform wall thickness throughout the part. Because of this, the ideal spot for gating is in the thickest area first. Then, the material can flow from thick to thin. As it begins to cool down and solidify, it has a better chance of filling when it travels from thick to thin. If that were reversed, the material would begin to solidify as it passed through the thin sections and not enough material would be available to finish filling the thicker sections. In addition, the part is much stronger when filling from thick to thin, as shown in Figure 5-10. The ball-shaped molecules traveling in the part gated from thick to thin are steadily compressed and bonded tightly throughout the part. However, in the part gated from thin to thick, the molecules are expanded and not well bonded, and the material has not been allowed to "pack." This results in a much weaker part. Our first desire then is to locate the gate in the thickest area of the part.

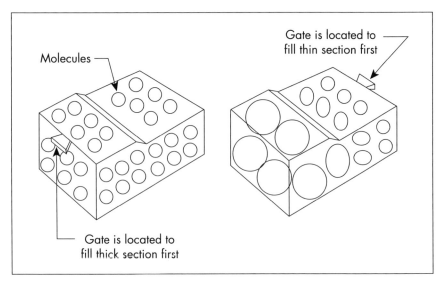

Figure 5-10. Gating to fill from thick to thin.

Next, we need to determine the dimensions of the gate being used. To do that, we begin by understanding the specific material we are molding. Easy-flowing materials can flow through thinner gates, and materials that are high in viscosity require thicker gates for proper flow. This *flowability* is determined by the *melt index* (MI) value of the specific plastic, which is measured according to the ASTM test method number D1238. Every plastic will have its own MI value, and these are published data available from the material supplier. The most common MI value is in the range of 12 to 14, but the overall range of MI values lie somewhere between 0 and around 50. The *lower* the number, the *stiffer* the material. So, a material with a MI of four would require a much larger gate opening than one with a MI value of 14. After the MI value is known and understood, the gate can be designed. The gate can be viewed as appearing like a window in a wall. It is an opening through which molten plastic will be flowing. The actual dimensioning of a gate depends on what gate design is being used. Some of the more common designs will be discussed in the following section. In the meantime, let us determine how many gates should be used.

The closer the cavity image is to the centrally located sprue bushing, the smaller the gate opening can be. This is important because the thickness of the gate is one factor that determines overall cycle time. As the gate gets thicker, cycle time increases. In addition, every gate must have a runner associated with it. This might be part of the main runner or a secondary runner feeding off the main runner. The more gates there are in a mold, the more runners there must be, and the farther away from the sprue bushing the cavity will be. This reinforces the concept that a part should be filled using a single gate, if at all possible. However, experience has shown that some part designs may warrant the addition of a gate if the primary gate results in a part that tends to warp or not fill properly. As a general rule-of-thumb, a gate should exist for every 8 to 10 in. (203 to 254 mm) of flow, in any direction. However, due to the differences in shear sensitivity, specific heat rates, and MI, it is better to analyze gating situations using one of the many finite analysis programs available today. If these are not available (or considered too expensive), experience and common sense must prevail. This may result in trial-and-error attempts, starting with a single gate and adding gates as necessary. A single gate is the most desirable. Depending on the material and basic part design feature, it is conceivable that *any* part can be molded with a single gate.

Sprue Gates

Sprue gating has long been considered the best way to gate a part. It is used primarily in single-cavity molds due to its basic nature, as shown in Figure 5-11. In this situation, the molten plastic is injected directly into the cavity image without the need for runner systems, and the gate is circular in cross section. Because the cavity is located centrally in relation to the sprue bushing, the flow of plastic is central to the cavity. This results in even flow distribution across the entire cavity. Stress is minimized by this condition, and flow lines are greatly reduced. In addition, sink marks are virtually nonexistent because the sprue stays molten all during the injection phase of the process, and holding pressure can be applied until the material in the part solidifies, thus minimizing shrink conditions.

Dimensioning the sprue gate requires that the diameter where the sprue meets the cavity be slightly larger than the wall thickness of the part at that junction. This ensures the sprue bushing will stay molten while the material in the cavity is flowing and solidifying. The shape of the sprue bushing, which is commercially available, forms the sprue. The internal taper is usually about .5 in./ft (42 mm/m), but is actually determined by the overall length required, as well as the major inside diameter already mentioned. The minor inside diameter must be the same as, or 1/32

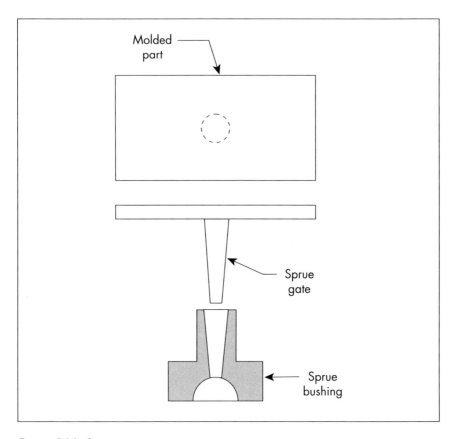

Figure 5-11. Sprue gate concept.

in. (0.79 mm) larger than, the nozzle of the molding machine. The nozzle orifice for the molding machine is interchangeable and determined by the viscosity of the material.

The biggest disadvantage in using sprue gates is that they must be removed from the molded part. Even the best removal process leaves some visual evidence of the sprue gate, and this may be objectionable on specific part designs. A decorative decal can be used to cover the vestige, or the sprue gate can be moved slightly off center, if necessary, to help disguise the condition.

Various Surface Gates

This section will describe a basic surface gate, edge gate, disc gate, and ring gate.

Basic Surface Gate

While a surface gate with a circular cross section is ideal for creating minimum stress, it is difficult to machine into a mold and keep concentric around its centerlines. This is because of the expansion and contraction movement of the steel over time. Any offset that occurs can create undue shear that may result in thermal degradation of the plastic. Therefore, rectangular surface gates have become the norm.

Figure 5-12 shows how a basic surface gate can be dimensioned. The basic surface gate is the most popular gate used. It is an opening in a wall and has three essentials: D (depth), W (width), and L (land). Each requires logical dimensioning that is based on the viscosity of the material being molded. Stiff materials (high viscosity) require larger openings than easier-flowing materials.

We will start with the D dimension. Remember that we want to gate into the thickest section of the part. The depth of our gate (D) will be 40 to 90% of that thickness. For instance, as shown in Figure 5-12, if the wall thickness is .10 in. (2.5 mm), the D dimension of our gate would be between .04 and .09 in. (1 and 2.2 mm). The thinner dimension would be used for easy flowing materials, such as nylon (a polyamide), while the thicker dimension would be used for stiffer materials, such as polycarbonate. We want the gate to be as thin as possible because it determines the overall cycle time (thicker gates require longer cooling

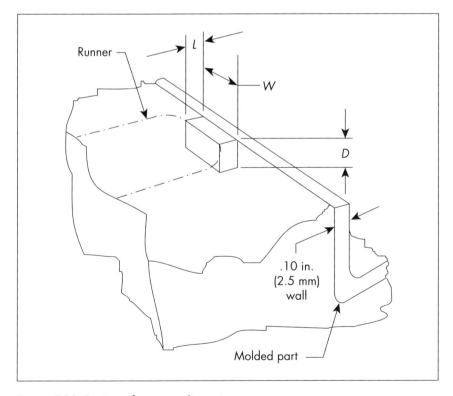

Figure 5-12. Basic surface gate dimensions.

time). So, we should start with the thinner dimension and only increase it if necessary. A good source of information for desired gate thickness is the material supplier of the specific material being molded. For our example, though, let us use 50% or exactly half of the wall thickness, which would be .05 in. (1.27 mm).

After the D dimension has been determined, we can develop the W (width) dimension. A rule-of-thumb states that the width of the gate should be at least double the depth, and can be as much as 10 times the depth. We can start with W being twice D. In our case that would be two × .05 in. or .10 inch (2 × 1.27 or 2.54 mm). We can increase W later, if needed.

That brings us to the L (land) dimension. Consider this the gate length. The L dimension should be half of the D dimension, but never greater than 1/16 in. (1.6 mm). If the L dimension is too great, it will cause the molten material to begin to solidify because the plastic must travel a long distance through a small opening. The surrounding steel pulls heat from the plastic quicker than if the L dimension is short. A typical effect of a land that is too great is "worming," a snakelike appearance on the molded part surface, emanating from the gate area.

Edge Gate

The edge gate is a variation of the basic surface gate and is used primarily for molding parts with large surfaces and thin walls, such as flat plates. Figure 5-13 shows a typical edge gate design. Note that the land (L) of this gate design is thin and narrow, and there is a secondary runner (d diameter) feeding from the primary runner. The narrow land acts as a throttle and causes molten material to fill the secondary runner area before it can enter the cavity image. This results in a uniform filling condition as the molten plastic leaves the land area and continues across the cavity image. The result is a part with uniform shrinkage in all directions (even crystalline materials) and parallel molecular orientation across the whole width, which is critical for optical products such as lenses.

Dimensioning an edge gate depends upon the viscosity of the material being molded, but, in general, the following values can be used. The L dimension should be from .02 to .06 in. (0.5 to 1.5 mm) and this land should run across the entire width of the part. The thickness of this land (L) should be 25 to 75% the thickness (t) of the part being molded, depending on how easily the plastic flows, with the stiffer plastic requiring the higher percentage.

The diameter of the secondary runner (d) should range in thickness from the same as the thickness (t) of the part being molded, to 30% greater than that thickness (t + 30% of t), depending on how easily the plastic flows. If d must be greater than t, the overall cycle time will be excessive because the d dimension will take longer to solidify than the part itself. Like the land, the secondary runner d should span the entire width of the part.

Disc Gate

Use of the disc gate design allows uniform filling when molding a cylindrical, sleeve-shaped part, as shown in Figure 5-14.

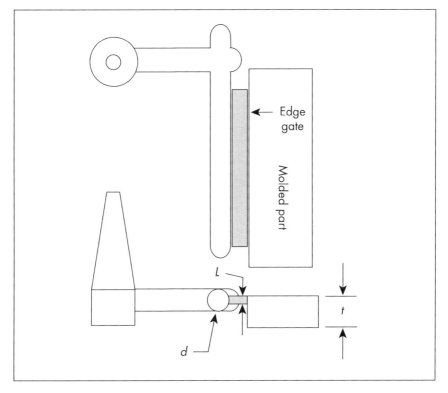

Figure 5-13. Typical edge gate design.

A disc gate is actually a diaphragm. It spans the opening at one end of a sleeve-shaped part and feeds material directly into the part. The diaphragm is thick in its primary body, equal to or 25% greater than the wall thickness of the molded part. This area, known as the land is directly connected to the molded part. The thickness (L) of this land should be 50% to 75% of the molded part wall thickness. The land should be .02 to .06 in. (0.5 to 1.5 mm) in width around its perimeter.

Ring Gate

A ring gate is a variation of the disc gate and is used for molding long, sleeve-shaped, cylindrical parts that need internal cores to be supported at both ends of the part. The ring shape results from the use of two runners, or feed channels. The main runner feeds a secondary runner that encircles (internally or externally) the cavity image, as shown in Figure 5-15.

External ring gates are the most common, and can be used if gate witness lines are visually acceptable. If not, internal ring gates may be used; however, they are more expensive to create and result in additional knit lines (at least two) being formed. The nature of ring gate designs results in knit lines. While these are

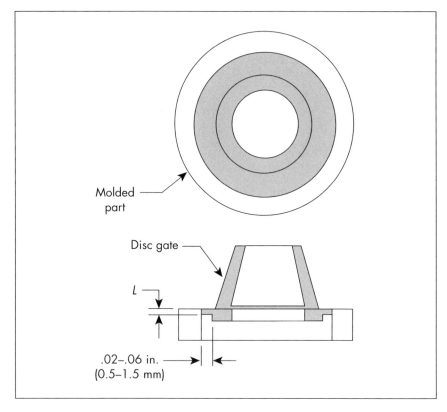

Figure 5-14. Disc gate design.

unavoidable, they can be minimized through processing parameters. Ring gate knit lines are always stronger than knit lines formed by other conditions.

As always, the dimensions of the ring gates depend on the viscosity of the material being molded. The *D* dimension of the primary and secondary runners should be 25% greater than the wall thickness of the part being molded. The thickness of the *L* (land) dimension should be 50 to 75% of the wall thickness of the molded part. The width of the land should be .02 to .06 in. (0.5 to 1.5 mm) around the perimeter of the cavity image.

Drop Gates (Three-plate Molds)

Sometimes called *pin gates*, drop gates are useful when a part cannot be gated with conventional surface gates due to aesthetics or mechanical interference of the gate vestige in later assembly. Drop gates allow automatic degating of the part from the runner as the mold opens and the parts are ejected, as shown in Figure 5-16. The drop gate is formed as the result of using a three-plate mold construction. (Also

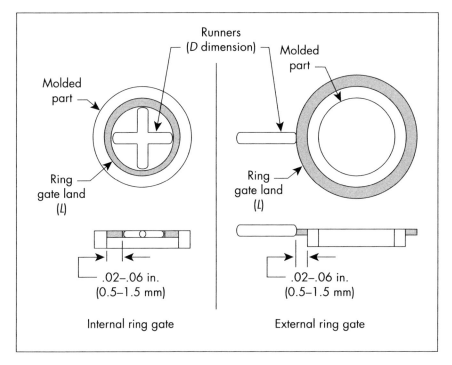

Figure 5-15. Ring gate designs, internal and external.

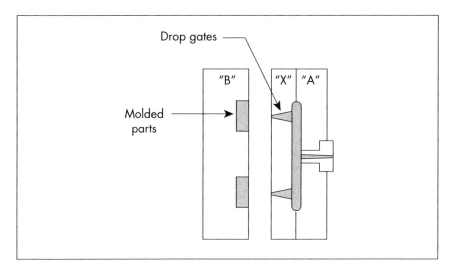

Figure 5-16. Drop gate (pin gate) concept.

see Figure 5-8, insulated runner concept.) The three-plate mold allows us to locate the runner system between the A plate and the X plate. Then, conical holes are machined from the runner to the face of the X plate. These are the *drop gates*, called that because they drop from the runner to the cavity image.

The use of drop gates requires the runner to have a larger diameter so that enough material can travel the extra length incurred. In addition, the drop gate major diameters are large to keep the molten plastic from solidifying before the cavity is filled. Therefore, the entire runner system contains much more material than a standard surface runner would contain, and this material only has value as regrind.

The biggest advantage to a drop gate is the ability to place it practically anywhere on the molded part. It is especially easy if the gate can be located on the surface of the part formed between the X and B plates. With creative manipulation, the gate can be located inside the part, under the part, or on the sidewalls of the part, almost as easily. The advantage of placing the gate anywhere allows much better control of flow into the cavity image, and can even simplify construction of certain molds. As mentioned earlier, drop gates allow automatic degating of the molded part from the runner system during the molds operational phases.

Tunnel Gates

Tunnel gates, also known as subgates, submarine gates, and banana gates, are used for both automatic degating of molded parts within the mold, and locating gates in areas not accessible by standard surface gating. They are commonly used for multiple cavity molds producing small parts, but also can be utilized for larger parts and single-cavity molds.

As shown in Figures 5-17 and 5-18, tunnel gates have two basic design options: pointed tunnel or truncated tunnel. The pointed tunnel provides a smaller, rounder orifice than the truncated version and results in faster cooling time for the gate. However, by *freezing off* (solidifying) earlier, the pointed version does not permit long holding pressure time that might be needed for close-tolerance or high-strength products.

Figure 5-19 shows a curved tunnel gate, commonly referred to as a *banana gate*. Banana gates are not common but can be used for low-depth parts made of softer plastics, or when the sharper corners of standard tunnel gates are undesirable (possibly causing stresses or shear degradation).

Figure 5-20 shows a tunnel gate being used to gate into an ejector pin. This concept allows gating into the inside surface (or B side) of a molded part (similar to the banana gate), which may be required for aesthetic purposes. Remember, if the part is thicker there than anywhere else, gate the thickest section whenever possible.

One note of importance concerning all gates, but especially tunnel-type gates: an ejector pin must be located as close as possible to the junction of the tunnel gate and the molded part. This is to ensure proper separation of the gate from the

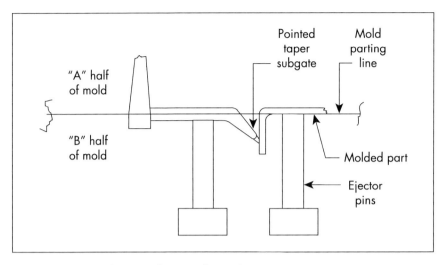

Figure 5-17. Tunnel gate with pointed tunnel.

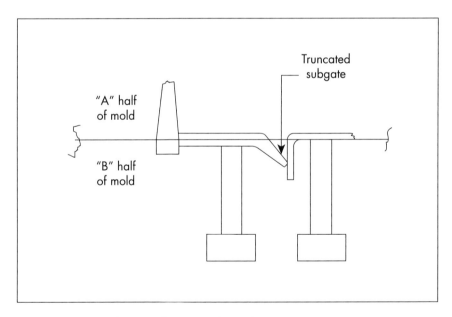

Figure 5-18. Tunnel gate with truncated tunnel.

part, and proper removal of the gate from the mold. Without this ejection, the gate may break away from both the part *and* the runner and stay in the mold causing successive cycles to produce blanks in that cavity.

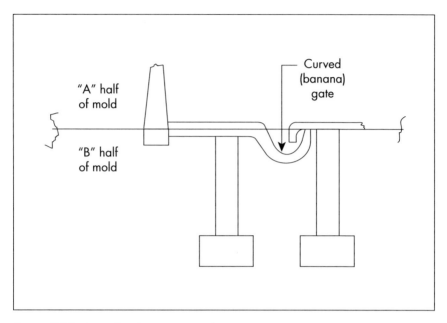

Figure 5-19. Curved or banana gate design.

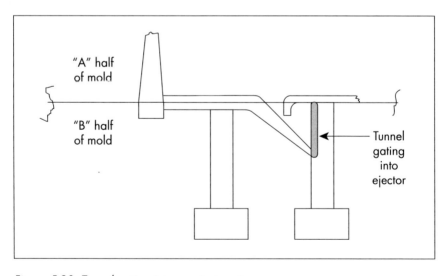

Figure 5-20. Tunnel gating into an ejector pin.

VENTING

This section will discuss the purpose, types, sizes, and locations of vents.

Why Vent?

Vents are needed to allow trapped air and processing gases to vacate the mold. When the mold is closed in preparation for injecting molten plastic, air is trapped in every cavity or opening within the closed mold, including the runner system. Without venting, this air will compress under the pressure of incoming material. In fact, it compresses the material so much that it will ignite and burn up the oxygen available to it. This results in charred plastic. It also requires very high injection pressure to overcome the resistance of the trapped air to compression. High injection pressures cause undue stress to be molded into the plastic part.

Types, Sizes, and Locations of Vents

Basic venting is a series of paths machined across the shut-off land of the cavity image, as shown Figure 5-21.

The vent has three dimensions: D for depth, W for width, and L for land. The vent can be thought of as a window in a wall. Plastic material fills the cavity image and pushes the trapped air toward the vent. The vent is machined such that it is thin enough to keep the molten plastic from traveling through, but thick enough to allow trapped air and gases to exit.

Dimensioning the vent begins with determining the proper D dimension. This can be taken from Table V-2 or estimated based on the viscosity of the flowing plastic. Easy flowing plastics, such as nylon, require a thin vent, while thicker vents can be used for high viscosity materials, such as polycarbonate.

Note that the vent depth for the runner as shown in Table V-2 is twice the depth of vent for the cavity. This is because we are not too concerned with potential flash on the runner, but flash on the part can cause injury to the end user, and loss of pressure buildup in the cavity. The column designating depth of vent for the cavity lists dimensions that allow air to escape but keeps plastic from entering the vent.

After the D dimension is determined, we need to determine the W dimension, or width of vent. The minimum dimension for vent width should be .125 in. (3.2 mm). A more practical and preferred dimension is .25 in. (6.4 mm). However, the W dimension has no maximum. In theory, the vent width can run all the way around the perimeter of the parting line of the cavity image, without stopping. Of course, it would then eliminate itself. Therefore, we need to be practical in determining the maximum W dimension. As shown in Figure 5-22, we use a rule-of-thumb that states that at least 30% of the perimeter of the cavity parting line should be vented. That leaves some strength to the steel surrounding the cavity image, but provides adequate venting.

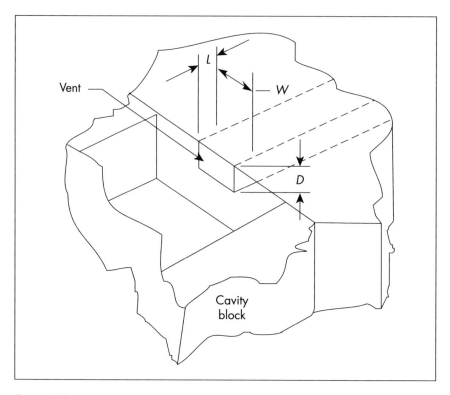

Figure 5-21. Basic venting concept.

In Figure 5-22, we show a cavity that has a total of 10 in. (254 mm) of perimeter. Using the rule-of-thumb mentioned, we would need a minimum of 30% of that 10 in. (254 mm) in vents. That would equal 3 in. (76.2 mm). If each vent was machined to the preferred 1/4 width, there would be a minimum total of 12 vents. If each vent was 1/2-in. (13-mm) wide, we would have a total of six vents, and so on. It is practical to assume that air will be trapped in routine areas, such as in each corner of the part, directly across from the gate, and in the area immediately surrounding the gate. Therefore, a good practice would be to locate vents at each corner, and place the remaining vents uniformly around the cavity. The more vents we can use, the faster the air will be removed, and the easier the molten plastic will enter the cavity (under lower pressure). GE Plastics (One Plastics Ave., Pittsfield, MA 01201) suggests that vents be placed at every 1 in. (25.4 mm) position along the entire perimeter of the cavity, as well as along the length of the runner.

Table V-2. Recommended Vent Depths

Material	Cavity	Runner
	in. (mm)	in. (mm)
ABS	.002 (0.05)	.004 (0.10)
Acetal	.0007 (0.017)	.0015 (0.038)
Acrylic	.002 (0.05)	.004 (0.10)
Cellulose acetate	.001 (0.025)	.002 (0.05)
Cellulose Acetate butyrate	.001 (0.025)	.002 (0.05)
Ionomer	.0007 (0.017)	.0015 (0.038)
Nylon 6/6	.0005 (0.0127)	.001 (0.025)
Polycarbonate	.002 (0.05)	.004 (0.10)
Polyethylene	.001 (0.025)	.002 (0.05)
Polypropylene	.001 (0.025)	.002 (0.05)
Polyphenylene oxide	.002 (0.05)	.004 (0.10)
Polyphenylene sulfide	.0005 (0.0127)	.001 (0.025)
Polysulfone	.001 (0.025)	.002 (0.05)
Polystyrene	.001 (0.025)	.002 (0.05)
Rigid PVC	.002 (0.05)	.004 (0.10)

We are now ready to determine the L dimension of the vent. Consider the vent, like the gate, to be a window in a wall. Air will be pushed through the window and allowed to escape to the outside atmosphere. The L dimension of that window determines how much buildup will occur due to condensation caused by the hot air passing through the relatively cool window opening. The farther the air travels, the more likely it will condense and form a deposit in the window opening. Therefore, the L dimension must be kept to a minimum, but no less than .031 in. (0.79 mm). Less than that may result in mold steel chipping out. For a maximum, we want no more than .125 in. (3.2 mm) for the L dimension. Any more than that will cause excessive condensation and deposits to be formed.

Removing trapped air from normally inaccessible areas requires other types of vents. For example, such vents can be used at the bottom of a blind hole, at the base of deep walls or corners, or where parting line vents are not effective. Figure 5-23 shows how these locations can be vented.

The vent can be machined onto the side of a movable (or stationary) pin. If the pin is movable, such as an ejector pin, the vent can be designed to be self-clean-

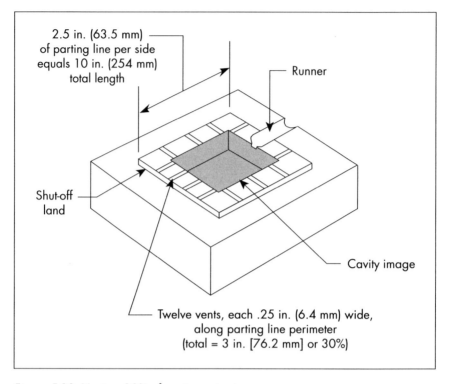

2.5 in. (63.5 mm) of parting line per side equals 10 in. (254 mm) total length

Runner

Shut-off land

Cavity image

Twelve vents, each .25 in. (6.4 mm) wide, along parting line perimeter (total = 3 in. [76.2 mm] or 30%)

Figure 5-22. Venting 30% of cavity perimeter.

ing and will remain "open" for a longer period of time before requiring a cleaning operation. The left section of Figure 5-23 shows such a self-cleaning vent on an ejector pin. Every time the ejector moves forward and returns, the vent area is cleaned. Nevertheless, even a stationary pin can be used effectively. The right section of Figure 5-23 shows such a pin.

Actually, any portion of the cavity that is laminated can act as a vent. If the fit of the laminated sections is such that they are airtight, vents can be machined onto the mating surfaces forming the fit. Alternatively, vents can be placed along the sides of slides and cams, as well as lifters and other action areas of a mold. These will provide sufficient venting until they become clogged, which may take a few thousand cycles to occur. Then, they can be easily cleaned by hand using a common nylon-scrubbing pad.

Any venting should be performed using accurate machining methods. The process of waiting until the first part is molded, looking for burned areas on the part, and placing a vent there on the mold using a hand-held, high-speed grinder, is obsolete. Due to the awareness that injection pressure must be accurately controlled, and stress must be minimized, venting must be considered a science and no longer

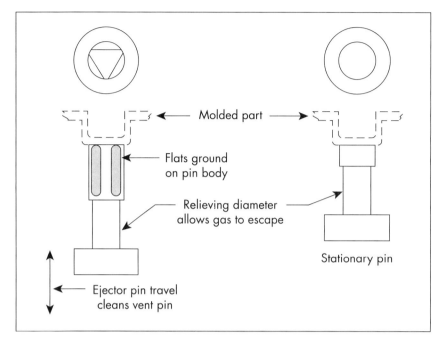

Figure 5-23. Venting nonparting line areas.

an art. Venting requirements can be determined accurately and before the mold is placed into production, thus reducing the time needed to "debug" the mold.

Remember, there is no such thing as too much venting. Even the 30% figure stated earlier must be considered a minimum value. Maximum venting (approximately 50 to 75% of the cavity perimeter) allows even faster removal of trapped air and easier injection of incoming plastic. Incorporating a vacuum system on the mold can assist venting. This consists of machining an O-ring (or other seal) groove around the parting line, placing a vacuum line within that seal, and timing the vacuum to start as the mold closes. The vacuum then pulls air from the cavity, while helping to pull incoming plastic into the cavity. Such systems are available commercially at an average cost of about $1,000 per mold. Venting the runner (at least every inch (25.4 mm) along the runner length) removes the air trapped in the runner system so it does not add to the volume of air already trapped in the cavity.

SUMMARY

There are three main items needed for the proper filling of plastic in an injection mold: a runner, a gate, and a vent.

The sprue bushing acts as an interface between the injection molding machine's cylinder nozzle and the runner system of the mold.

Runner systems are utilized to direct the flow of the molten plastic, after it enters the mold, from the sprue bushing to the cavity images that are used to form the molded product.

Surface runners are the most common and require a round cross-sectional area to create equal pressure in all directions. This minimizes molded-in stress.

Highly viscous (very stiff) materials require larger-diameter runners than low-viscosity materials.

Right-angled turns in a runner system require an additional 20% increase in the diameter to compensate for pressure drops. A more direct path, such as a wheel spoke design or curved runner path, is preferred to using right-angled turns.

Insulated runners evolved from three-plate mold designs. They eliminate regrind and scrap buildup caused by standard surface runner systems.

Hot runner systems evolved from insulated runners and, in effect, move the nozzle of the molding machine right up to the cavity image, eliminating runners and sprues.

A part should be gated into its thickest section, from thick to thin, never the reverse.

Cavity sets should be located as close to the sprue as possible to minimize travel time and distance.

Subsurface gates require an ejector pin be placed as close to the junction of the gate and cavity image as possible.

Venting is required to allow trapped air and processing gases to vacate the mold.

Proper venting requires that a minimum of 30% of the parting line be vented. There is no such thing as too much venting. Both the cavity set and the runner need venting.

QUESTIONS

1. What three items are required in the mold to ensure proper filling of the molten plastic from the machine's heating cylinder to the cavity image?
2. What will happen if the sprue-bushing radius is smaller than the matching radius of the machine nozzle?
3. What is the purpose of the cold slug well?
4. Why is a round cross section for a runner the best design?
5. What is the ideal runner diameter for acrylic flowing through a three-inch long runner?
6. How much does a runner diameter need to be increased when making a right-angled turn?
7. Why is a "wagon-wheel runner layout" the best design?
8. What are the three major advantages to using a hot runner system?
9. Where should the gate be located on a part with varying wall thickness?
10. What standard test determines the flowability of a molten plastic?

Controlling Mold Temperatures

6

OVERVIEW

After the molten plastic is injected into the cavity image, it is allowed to stay there, under pressure, until it has cooled down and solidified enough to be removed from the mold. The plastic does not need to be totally cooled, just enough to allow ejection of the finished product without unacceptable distortion occurring to the plastic. This cooling is done by a mold temperature control system that removes heat from the mold and maintains the correct mold temperature. In the following section, we will examine the more common methods used to accomplish this feat.

WATERLINES

The use of waterlines machined throughout the mold to allow water to flow through the mold is the most common method of controlling mold temperature. This is accomplished by drilling holes (1/8-, 1/4-, and 3/8-inch pipe are the most common) as close as possible to the actual molding area of the cavity sets (see Figure 6-1).

Because of the accuracy required, the drilled holes are usually machined using a gun drill or deep-boring tool. Then, they are fitted with pipe threads for connecting nipples, hoses, and/or quick-disconnect fittings. Hoses are connected to floor-mounted temperature-control units or to machine-mounted water manifolds. In all cases, the temperature difference should not be more than 10° F (5.5° C) between any two points. This includes on the cavity molding surfaces, on the same mold half, or on both mold halves. A difference of more than 10° F (5.5° C) creates excessive stress in the part and will result in unbalanced plastic flow during the injection phase. These areas also shrink more than the cooler sections.

Control Units versus Manifolds

There are two primary ways to control mold temperatures: through use of control units and manifolds. Though the methods are different, the results are the same.

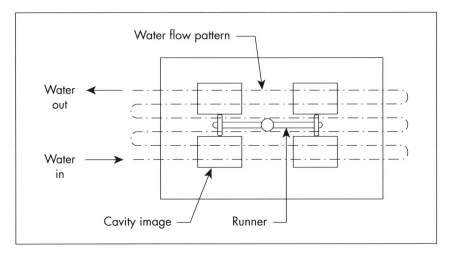

Figure 6-1. Common waterline layout.

Control Units

Figure 6-2 shows a mold temperature control unit. If floor-mounted temperature control units are used, a single unit should be used for each half of the mold. This eliminates the practice of using heated water from the first mold half to "cool" the second mold half, as is done when only one control unit is used.

The unit has a set point that is determined by the user. Water is circulated through the unit and heated until it meets the desired set point. As the water recirculates and the temperature increases, the unit begins to replace the heated water with tap water to maintain the desired temperature. This process continues as the unit functions to maintain the desired temperature. The circulating water temperature is displayed through a thermometer mounted on the unit. Remember that this is only an indication of the water temperature and *not* the mold temperature. The mold temperature must always be measured by using a surface pyrometer on the cavity surfaces.

Manifolds

If water manifolds are utilized, the water flow is manipulated by throttling shut-off valves at the connection points of the manifold. Each line can be controlled independently. The mold temperature is determined by the flow of the water through these valves, but must be measured by a surface pyrometer on the mold surface itself. In most cases, the manifolds are connected to a source of chilled water, usually at a temperature of approximately 50° F (10° C). The manifold concept is based on a system that can control the mold temperature after it is established, but requires a slow and steady buildup as the metal of the mold absorbs heat from the injected plastic. The manifold system then depends on the

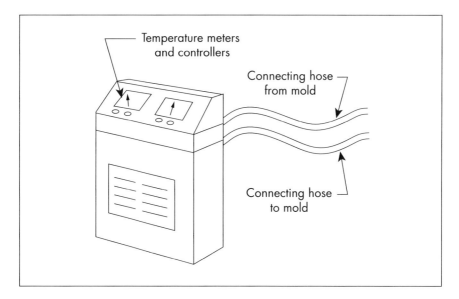

Figure 6-2. Mold temperature control unit.

continuous heat of incoming plastic to create a buildup of heat in the mold to provide the proper molding temperature. After that temperature is reached, the manifold system can maintain it. However, this process may take an hour or two to level out and the molding parameters may have to be adjusted constantly while this is happening to ensure properly molded parts. This creates a condition of having to "tweak" the manifolds during the first hour or so of production until the mold reaches the intended temperature.

Some molders use a colder mold for faster cycles and more profits. In fact, most parts require a slow cool down period especially when molding crystalline materials. This normally means a warm mold and long cooling cycles to create the highest level of physical strength in the final part. Cold molds will negatively affect the physical properties of a molded part, while warm molds will enhance those same properties. There are few situations where cold molds should be used. If manifold systems are in place, there is a tendency for the molder to routinely lower the mold temperature to the temperature of the manifold system, which is usually around 50° F (10° C). This may result in faster cycles, but can create a major quality problem if not understood and controlled.

Laminar Flow versus Turbulent Flow

There are two different types of flow that water can experience when traveling through a waterline of a mold: laminar or turbulent. Figure 6-3 shows the differences between the two conditions. Both conditions will remove heat from the

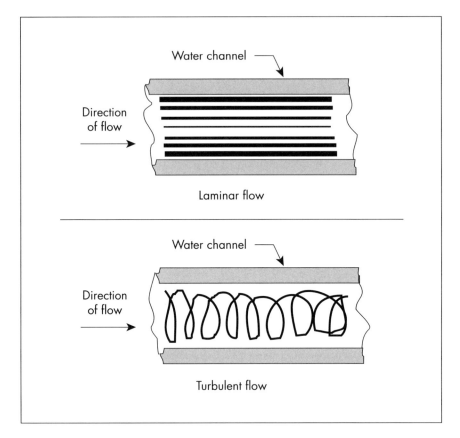

Figure 6-3. Laminar flow versus turbulent flow.

surrounding mold metal, but the laminar flow is not nearly as effective as the turbulent flow. Note that in the laminar flow diagram the water travels in separate layers. The layers nearest the outside are next to the mold metal and are in direct contact with the heat that needs to be removed. These layers move slowly (due to friction) and transfer some of that heat to the faster-moving inner layers. However, the very center layer, moving fastest of all, receives no heat at all. In the turbulent flow model, the water is constantly being tumbled and mixed. All of the water is in contact with the mold metal at one time or another and all of it is used to remove heat from the mold metal. This is the desired effect.

The creation of turbulence is a function of flow rate, waterline diameter, water viscosity, water temperature, and velocity of the water as it travels through the channels. Whether concerning laminar or turbulent flow, these conditions are characterized by a ratio known as the *Reynolds number*. Conditions causing a Reynolds number of 2,000 or less will result in laminar flow. Ideal turbulence is found when conditions create a Reynolds number of 3,500 or more. In between exists a transition area that fluctuates between laminar and turbulent flow.

Determining the existing Reynolds number can be achieved by using the following formula:

$$R = KQ/Dn \qquad (1)$$

where:

K = 3,160
Q = flow rate (gpm)
D = diameter of waterline (inches)
n = water viscosity (centistokes), see Table VI-1

The most important answer we are attempting to find is the gallons per minute (gpm) required to produce a specific Reynolds number. The gpm is easily variable when all other conditions are fixed. So let's perform an exercise to make that determination.

Table VI-1. Water Viscosity versus Temperature

Water Temperature (° F [° C])	Viscosity (n) (centistokes)
32 (0)	1.79
50 (10)	1.30
68.4 (20.2)	1.00
100 (37.8)	.68
150 (65.6)	.43
212 (100)	.28

First, we must know the temperature of the water entering the mold. We'll use the temperature of 50° F (10° C). Then, we need to know the diameter of the waterline we are using. If we use a 1/8-in. (3.2-mm) pipe diameter, the actual drilled hole-opening diameter (water passage) is 11/32 in. (8.7 mm). Because we want to obtain turbulent flow, we will use a specific minimum Reynolds number of 3,500. Remember, our formula is: $R = KQ/Dn$. By substituting the numbers for the variables, we get:

$$3,500 = (K \times Q)/(D \times n) \qquad (2)$$

where:

K = 3,160
Q = variable to solve
D = 11/32 or .34375
n = 1.30

So,

$$3,500 = (3,160 \times Q)/(.34375 \times 1.30)$$
$$Q = (3,500 \times .34375 \times 1.30)/3,160$$
$$Q = .495 \text{ gallons per minute (gpm)}$$

Therefore, a flow rate of .495 gpm with water temperature of 50° F (10° C) will create proper turbulence (3,500) through a 1/8-in. (3.2-mm) pipe waterline. The formula can be adjusted for determining what Reynolds numbers are used for specified waterline diameters and flow rates, or what water temperature would be needed to create a specific Reynolds number. There are variable conditions under which the ideal Reynolds number range (3,500 to 7,000) can be achieved. Strive to maintain a Reynolds number of 5,000, if possible, to ensure that proper heat transferability is provided for the coolant that controls the operating temperature of the mold.

You can easily detect whether or not a mold temperature is being properly maintained by noting the temperature difference between water going in and water coming out. Contrary to popular belief, there should not be more than a 10° F (5.5° C) difference between the two temperatures. If outgoing water is hotter than incoming water, it means that there is too much heat retained in the mold and the water is not bringing it out fast enough. An ideal condition is one where the heat is removed as fast as it is created, which would result in the water temperature being exactly the same going in as coming out. While this may not be entirely possible, there should be no more than a 10° F (5.5° C) difference. If the water-lines have been designed for the proper Reynolds number and the return water-line is hotter, scale buildup may be occurring in the lines or some other item is plugging the flow.

Determining Location of Waterlines

An easy statement to make is that the waterlines should be located as close as possible to the surface of the cavity image forming the molded product. While that is easy to state, it is not easily accomplished. The reason is that drilled waterlines must follow straight paths, but most molded products have three-dimensional qualities and are not flat and straight. In many cases, drilled lines are placed so that they surround the part as much as possible, but do not take the exact configuration, as shown in Figure 6-4. The round-shape cavity is surrounded by a square-shape waterline pattern. This is not efficient and causes uneven cooling in the molded part. This is due to the uneven location of the water being used to pull heat from the plastic. The uneven cooling will result in a tendency for the flat part to warp and bow as some areas cool down quickly while others cool at a slower rate.

Figure 6-5 shows a more accepted waterline pattern for this product. The coolant follows a pattern that is much closer to the actual shape of the product being molded. However, to create this pattern, a complicated system of drilled and plugged waterlines must be created. This is expensive, and still does not form an

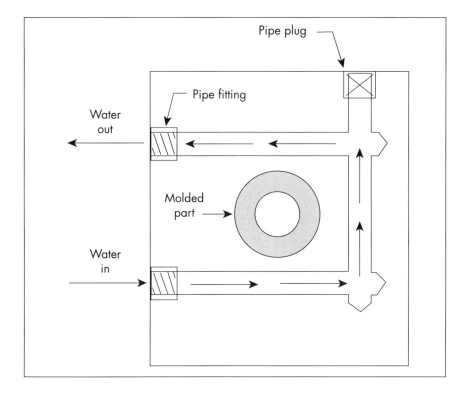

Figure 6-4. Improper waterline pattern.

ideal pattern. A better pattern may be the one shown in Figure 6-6, where the water flows in a pattern that very closely matches the shape of the product being molded. However, to incorporate this pattern requires a system of open-faced channels, connected together, and sealed with O-rings to eliminate leakage. While this provides the most acceptable pattern, it is expensive to create and requires constant maintenance to keep leaks from forming and damaging the molded parts, as well as the mold itself.

A further problem is that the steel used for making the mold must contain the high pressure initiated by the injection phase of the molding process. Therefore, the waterlines cannot be too close to the cavity or they will create a breakthrough of the cavity steel. A rule-of-thumb suggests that waterlines be no closer than 1.5 times their diameter, but a safer rule-of-thumb states they should be no closer than a full two diameters from the cavity, as shown in Figure 6-7. The diameter is determined by what is required to provide the proper Reynolds number value. However, that does not mean there should only be a single waterline. If depth allows, additional layers of waterlines can be used.

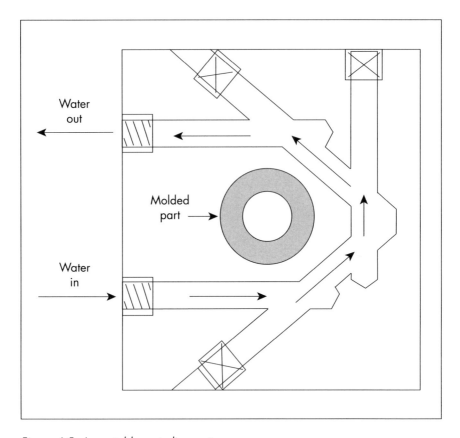

Figure 6-5. Acceptable waterline pattern.

When locating waterlines, other items that make up the construction of the mold may cause interference with the waterlines. For example, bolts are used to hold the cavity blocks in place. These come from behind the cavity blocks and might interfere with any waterlines running under these blocks. Also, ejector pins for the part and the runner must travel through the B side of the mold, and these also might interfere with waterline locations, as shown in Figure 6-8. It is critical that the mold designer lay these items out with two primary thoughts in mind: first, locate the cavity blocks as close as possible to the center of the mold (to minimize flow travel of the incoming plastic); and second, locate waterlines as close as possible to the contour of the cavity image. These two concepts must be brought to bear on each other until a compromise is created that satisfies both requirements.

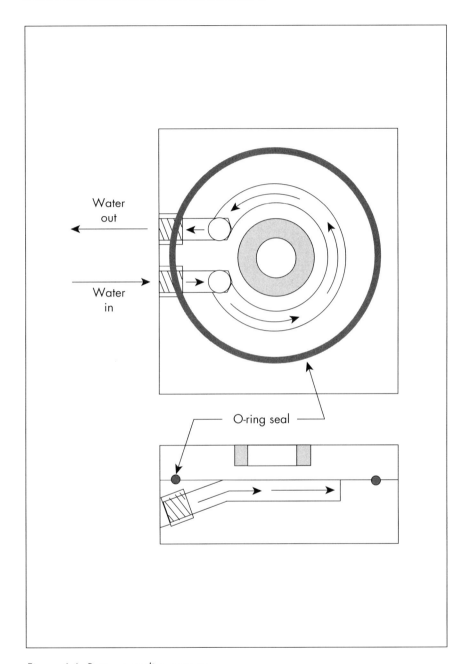

Figure 6-6. Better waterline pattern.

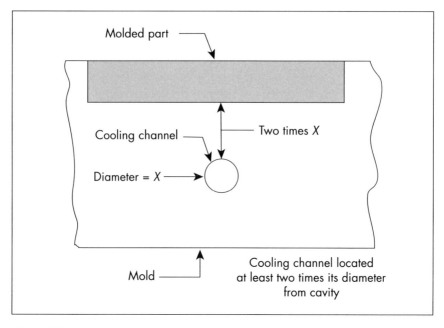

Figure 6-7. Waterline distance to cavity.

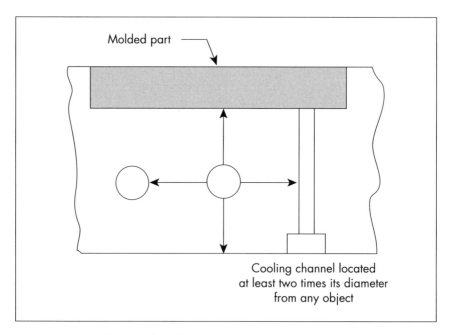

Figure 6-8. Waterlines and interference items.

Bubblers and Cascades

Sometimes it is difficult to get temperature control water located where it is needed, for example, the center of a deep metal core for making wastebaskets. In those cases, specially designed components can be used. One popular component is called a *cascade* and is commonly referred to as a *bubbler,* or sometimes a *fountain*, as shown in Figure 6-9.

In a bubbler, the cooling medium (usually water) comes from the main cooling channel, enters at the bottom of the bubbler, flows up through an inner tubular device, cascades inside the unit, and flows down through an outer tubular device, exiting back into the main cooling channel. This exiting water carries heat from the cascade unit into the main cooling medium.

Cooling Pins

Another device is known as a *cooling pin* (sometimes called a *heat pin or thermal pin*), as shown in Figure 6-10. This unit works on the conduction principle and is

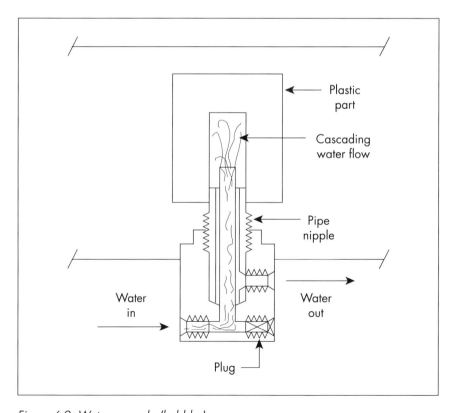

Figure 6-9. Water cascade (bubbler).

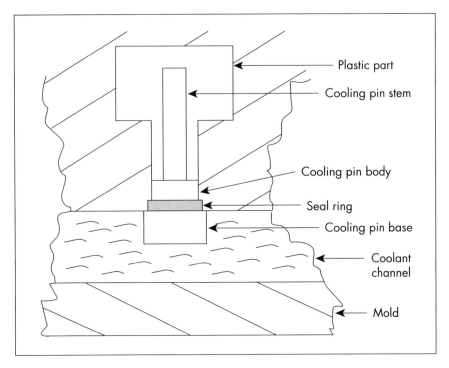

Figure 6-10. Cooling pin.

made from a thermally conductive material, such as beryllium copper. In Figure 6-10, the cooling pin is connected to the metal molding surface and the base of the pin sits in the main cooling channel. Heat is transferred from the plastic to the highly-conductive cooling pin. The cooling medium takes heat away from the cooling pin by removing it from the base of the pin where the heat has been conducted. In some designs, the pin is hollow and contains a liquid or gas that helps conductivity.

HEAT TRANSFER METALS

The molding industry uses a variety of metals for molds to assist in maintaining proper mold temperature. These metals are primarily aluminum and copper alloys. The ability of these metals to conduct heat allows them to be used in areas such as deep cores where it may be impossible to drill standard waterlines. In some cases, even thin metal foils or thin-walled tubing can be strategically located to pull heat from a cavity section.

All of the materials mentioned in the following sections have varying degrees of thermal expansion and none are identical to steel. In some cases, this may cause a problem for determining expansion rates between the mold steel and the component made from another metal. Some final fitting may be required after the mold is brought to proper operating temperatures to accommodate these differences, but in most cases, standard machining practices do allow room for this phenomenon.

Aluminum

Various aluminum alloys are available for injection mold components, with most being of a "wrought" condition, which results in very strong alloys. These aluminum materials possess a heat conductivity value that is approximately four times that of common mold steels, and results in an ability to pull heat from a molded part up to four times faster. For this reason, aluminum is a common material to use in areas where standard cooling channel design and location are not adequate to provide proper heat dissipation. Examples would include deep, cylindrical parts, parts with multiple, complicated parting lines, or parts requiring long, thin core pins. In these cases, it may not be possible to get any cooling at all using common cooling methods. Aluminum can be used for the hard-to-cool portion of the mold, as shown in Figure 6-11.

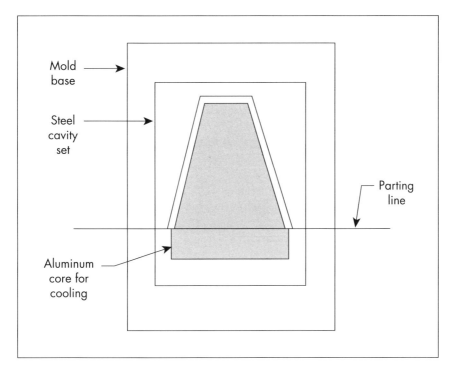

Figure 6-11. Use of aluminum for hard-to-cool portion of mold.

For most situations, the tensile and compressive strengths of aluminum are adequate, and the surface can withstand a high degree of wear in the normal state. If wear is a concern, the surface can be treated with a variety of methods, such as anodizing, to give a surface hardness of up to 65 R_c (Rockwell C).

Copper Alloys

The surface hardness of copper alloys (bronze) can be increased from 35 to 43 R_c. Thermal conductivity can be higher than aluminum, and bronze (specifically beryllium copper) has a rate of five to six times that of common mold steels. This can result in cooling time reductions of up to 40% when copper alloys are used for the entire cavity image. Copper alloys are ideal for hard-to-cool areas of steel molds and core pins (as well as other core shapes), and are perfect candidates for mold components used to reduce thermal differences in complete mold sets.

Zinc Alloys

In some cases, zinc alloys have been used for making mold components, such as core pins, needed to provide cooling where standard methods are not possible. However, surface wear is a problem and usually components do not fare well when total cycles number more than a few hundred. The thermal conductivity, however, is two to three times that of steel, so zinc alloys may be considered, especially in cases where an unusual shape must be used. The mold can be cast in zinc alloy and finish-machined to shape.

AIR COOLING

It is possible to use compressed air to perform the cooling action needed for injection molding. While this method is not as efficient as standard cooling methods, it is possible to use in situations where standard cooling is not practical or possible. For example, if there is a cracked waterline that causes pressurized water to leak into a cavity, air could be used instead of water. This may be suitable for finishing a production run before removing the mold and repairing it. Another case might be one where specific components, such as long-deep cores, are cooled by airflow instead of water, to equalize mold temperatures without relying on sophisticated water control devices.

SUMMARY

The mold temperature can be controlled through use of a temperature control system that is used to remove heat from the mold and maintain the correct mold temperature for the specific plastic material and product design being employed.

The use of waterlines machined throughout the mold to allow water to flow through the mold is the most common method of controlling mold temperature.

Temperature control unit settings determine the actual water temperature used to dissipate heat from the mold. The actual *mold* temperature must be determined by measurement, usually through use of a surface pyrometer placed on the mold metal surface.

Turbulent water flow is three to five times more efficient in controlling mold temperature than laminar flow.

The Reynolds number is used as an indicator of turbulence with 3,500 being minimum and 5,000 being ideal.

There should be no more than a 10° F (5.5° C) difference between any two points of a mold when measured in the cavity image.

Waterlines should be located as close as possible to the molding surfaces of the cavity.

Heat transfer metals, such as aluminum, copper, and zinc, have two to six times the thermal conductivity value of common mold steels and can be used to help control mold temperatures evenly.

Air can be used as a cooling medium in molds, but it is not nearly as efficient as water.

QUESTIONS

1. What is the most common method used for controlling mold temperatures?
2. What is the ideal maximum temperature difference measured between any two points of a mold?
3. What is the result of using one control unit to control the temperature of two mold halves?
4. If a temperature control unit gage is set at 150° F (65.5° C), what is the actual temperature of the mold?
5. Why is turbulent flow desired over laminar flow in a waterline?
6. What is the minimum Reynolds number to be used in determining turbulent flow?
7. What is the correct temperature difference that should be found when measuring the temperature of water coming out of a mold versus the temperature of the water going in?
8. How should waterline location be determined?
9. How does the thermal conductivity rating of copper alloys compare to common mold steels?

Mold Alignment Concepts 7

PURPOSE OF ALIGNMENT

The injection molding process is one that demands close control over a variety of parameters to achieve high quality products at the most efficient cost. One of the results of this is the requirement for precise alignment of the mold components to each other and of the mold to the molding machine. In addition, all of the molding machine components must be aligned to each other. Without proper alignment, the mold halves might not line up properly and may cause deviations in wall thickness and improper (and inconsistent) dimensions of the molded part. *Alignment*, for these purposes, can be defined as the accurate locating of components to their respective positions. This is accomplished in a variety of ways, and these methods will be discussed in detail in the following sections.

ALIGNMENT OF MOLD HALVES

Commonly, the prime method of aligning the two mold halves (A and B) is to use leader pins and bushings. There are a variety of designs to incorporate the concept, but the most common is the standard shouldered leader pin (with or without oil grooves), and a standard corresponding bushing. The shoulder design allows for a single boring tool to machine holes in both mold halves for precise location of the pin and bushing. Another popular method of mold half alignment is with tapered locks, commonly called *interlocks* or *parting line locks*. We will investigate both methods in the following section.

Leader Pins and Bushings

The purpose of the leader pin and bushing is to make sure the mold halves are properly aligned before the mold's moving components engage each other, and to ensure the mold is properly aligned before it fully closes. Therefore, the leader pin length and the amount of engagement into the respective bushing are important.

Figure 7-1 shows an example of typical leader pin and bushing components. Note the tolerances of the pin and bushing diameters. These are held closely to ensure precise alignment of the mold halves during closure. Figure 7-2 shows how the pin and bushing interact to achieve this alignment. The holes in which the pins and bushings reside must be jig ground for precise fit and location. The leader pin units must be located in each corner of the mold, and one unit should

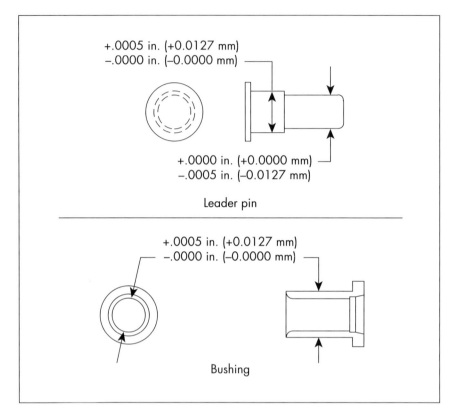

Figure 7-1. Typical leader pin and bushing components.

be intentionally offset to ensure that the two mold halves can only go together one way, as shown in Figure 7-3.

Leader pins should be placed as close to the mold edge as possible to allow maximum area for cavity images and waterlines. Leader pin dimensions are determined by the size of the mold being used, with the average diameter range being a nominal .75 to 1.5 in. (19 to 38 mm). However, pins are available for large molds in diameters up to 3 in. (76 mm). The length of the pin is determined by mold plate thickness and the overall location of the two mold halves at a point just before engagement.

Bushing diameters are adjusted to the appropriate pin diameters, and the bushing length should be one and a half to three times the inside diameter of the bushing. This allows for proper engagement.

Leader pins and bushings are made of case hardened steel with a hardness of 60-62 R_c. Wear can be minimized by lubricating with a molybdenum disulfide grease. Bushings (or pins) are available with oil grooves to contain such lubrication.

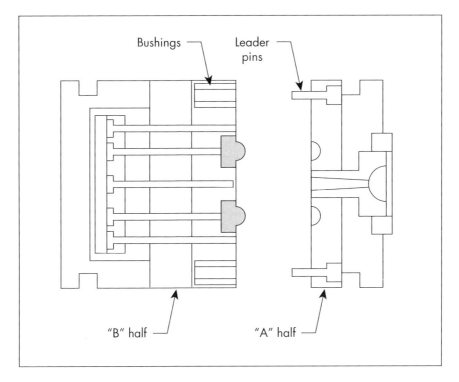

Figure 7-2. Engagement of leader pin and bushing.

Proper leader pins and bushings are usually included when standard mold base sets are purchased (from companies such as DME, 29111 Stephenson Highway, Madison Heights, MI 48071, a company which specializes in providing mold bases for moldmakers to finish).

Tapered Locks

The most common method of taper lock usage is on the primary A-B parting line. A conical post is mounted on one half of the mold, and this must mate with a conical depression on the other half of the mold. Such a system is shown in Figure 7-4.

Taper lock systems are available commercially from most of the mold base supply houses. However, they also can be manufactured at the time that the mold is made. The material is usually shock resistant steel, such as S-7, and the hardness usually ranges between 50 and 60 R_c (Rockwell C scale). They usually incorporate a 10° angle per side to attain the required taper design, but this can be as much as 20° per side. There can be as few as two sets per mold or as many as six or more sets, depending on the size of the mold. Taper lock systems are also available in rectangular shapes.

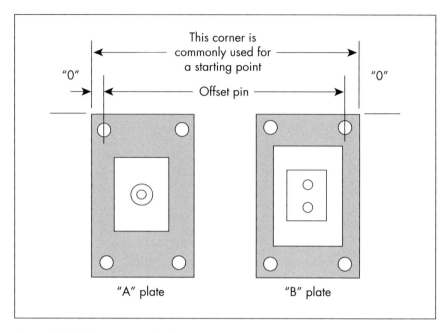

Figure 7-3. Offsetting one leader pin unit.

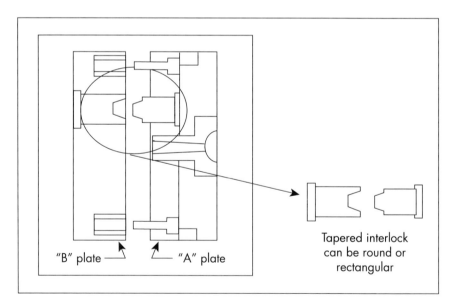

Figure 7-4. Typical taper lock system.

Another type of parting line interlock, shown in Figure 7-5, is mounted on the outside surface of the mold at the parting line, and is sometimes called a *straight side interlock*. Straight side interlocks are designed to provide positive alignment for molds with interlocking cavity and core sets. They furnish exact location and can be used in conjunction with conical-shaped taper locks (or rectangular-shaped taper locks) for extreme situations.

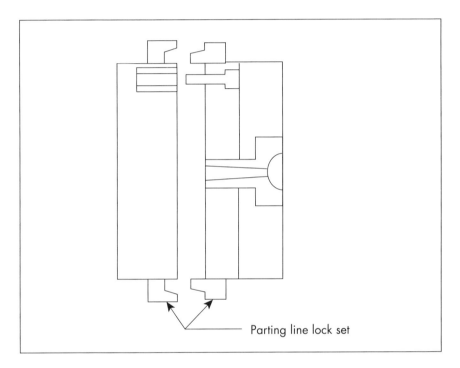

Parting line lock set

Figure 7-5. Straight side interlock.

ALIGNMENT OF MOLD COMPONENTS

This section will discuss typical action lock mechanisms, slides, cams, lifter systems, and telescoping components.

Typical Action Lock Mechanisms

Some product designs require moving components within the mold's interior. Such components include slides, cams, and lifters, all of which are called *actions*. These actions separate portions of the mold on either the A half and B half side, or both, when the mold opens, and they must be aligned for precise location as they are positioned to return when the mold closes. This is shown in Figures 7-6, 7-7, and 7-8.

Slides

Slide mechanisms are large areas of the mold that are pulled away from the main cavity sections by any of many methods. The most common is the angled pin system. As shown in Figure 7-6, the angled pin is stationary (on the A half of this drawing), and the slide mechanism has a hole in it that matches the angle of the stationary pin. When the mold opens, the slide is forced to follow a path pulling it away from the main cavity section due to the angled pin and hole combination. When the mold closes, the slide is pushed forward by the relationship between the stationary pin and the angled hole until the mold is fully closed and the slide

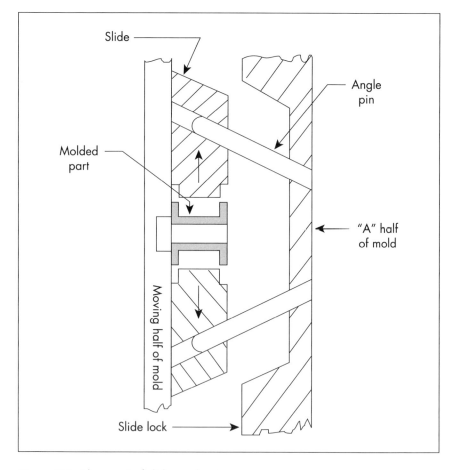

Figure 7-6. Alignment of slide mechanism.

is fully forward. However, due to the side forces from the injection pressure used in the molding process, the slide actions are pushed slightly away from the cavity sections. This causes flash and dimensional discrepancies to occur on the molded part. To overcome those side forces, locking mechanisms are usually incorporated. These are identified in the drawing as nothing more than large wedge-shaped blocks of steel mounted on the A half. The wedging action is attained by using an angle of approximately 15° (up to 25°) on the mating faces of the lock and the slide. The lock mechanism is sometimes referred to as a *wedge lock* or *heel lock*.

Large, hydraulic cylinders can provide the locking force instead of the wedge lock device. However, hydraulics are not as dependable as mechanical wedge lock systems and must be precisely electronically synchronized as the mold opens and closes.

Cams

Cam mechanisms are similar to slides, but they usually require only a small area of the movable portion of the mold. For example, cams can be incorporated when using a core pin through a sidewall or to have portions of the molded product form an undercut, as shown in Figure 7-7.

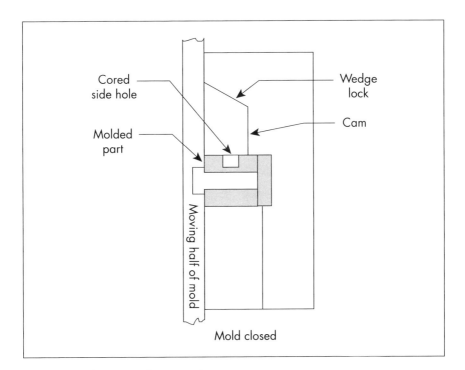

Figure 7-7. Alignment of cam mechanism.

As with other systems that result in parts of the mold separating when the mold opens, cam systems must be repositioned properly and accurately when the mold closes. The cam can be considered a small slide and locked in place using a small wedge lock device. Alternatively, a small hydraulic cylinder can be fitted to provide constant hydraulic pressure against the cam during the molding process. This must be timed to operate precisely, and is not as dependable as strictly mechanical systems.

If the cam is operating a core pin, that pin must be placed within a bushing device that allows perfect alignment every time the mold opens and closes. A pin that is not aligned will not only wear unevenly and possibly gall, but will produce a molded part with inconsistent dimensions in the area formed by the cam.

Lifter System

A lifter is commonly used in forming an internal hole, or other undercut situation, on the inside wall of a molded part. It consists of an area of steel that must be moved out of the way as the mold opens to eject the molded part. It must be moved back into its original position as the mold closes to form the undercut it was designed to form. Of course, to be located properly and accurately, it must be aligned. To be held in position during the injection phase of the molding process it must be locked in place. These activities are accomplished as shown in Figure 7-8.

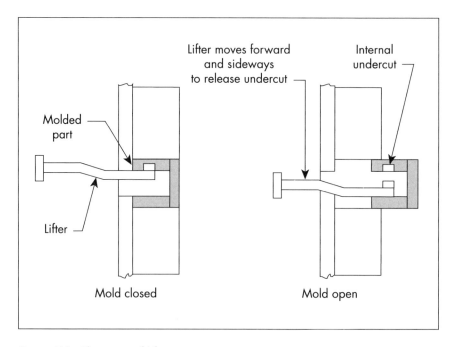

Figure 7-8. Alignment of lifter system.

Telescoping Components

In molding a part with a long core pin, or similar feature, it is desirable to support it during the injection phase of the molding process or it may be deformed by the injection pressures directed against it. A good example of this problem occurs in the molding of ballpoint pen barrels. Figure 7-9 shows the mold components for such a product.

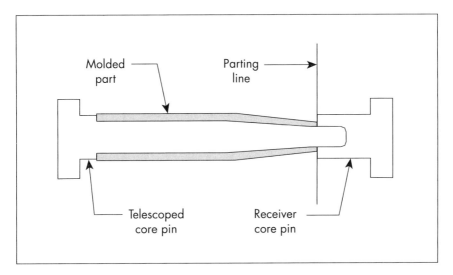

Figure 7-9. Telescoping core pins.

One method of providing support for the long core pin is to force it to engage with another component at the far end of the pin. This is referred to as *telescoping*. This method consists of positioning the core pin while the mold is closed. During this time, the tip of the core pin enters a holding device that provides support for the core pin until the injection process is complete. When the mold opens, the core pin disengages from the holding device to allow for ejection of the finished product.

Telescoping can be used for core-to-cavity alignment too. Alternatively, it can be utilized in any situation requiring support and/or alignment of any mold component. The critical factor is to design and build the telescoping devices to proper tolerances and clearances to ensure adequate alignment without allowing material to flash between the telescoping components. Telescoping requires not only accurate machining methods, but proper cooling in the entire mold. This is so that there is consistent expansion and contraction of all mold components and, consequently, minimized possibility of interference situations created by the thermal expansion differences of the components.

Many mold designers insist on using telescoping core pins in all situations requiring molded-in holes, even if the core pin is short relative to its length. The reason lies with the normal thermal expansion incurred during the molding process. When the mold heats up and expands, the core pin length may not be long enough to continue mating if a pin is made to fit perfectly against the mating half of a mold. Flash may form across the hole, as shown in Figure 7-10. To solve this problem, most moldmakers will "preload" the pin by building it slightly longer than actually needed (by .001 in. [0.025 mm], or so). Then, as the mold expands, the preload condition is enough to keep flash from forming. However, preloaded pins can cause a "mushrooming" of the pin face that will cause an undercut situation and result in problems in removing the molded part from the mold. Telescoping the core pin into the other half of the mold eliminates this problem, as shown in Figure 7-10.

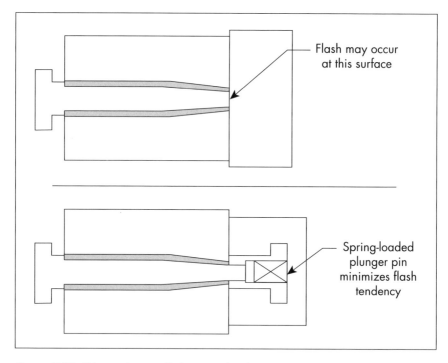

Figure 7-10. Telescoping to eliminate preload.

ALIGNMENT OF MOLD TO MACHINE

The mold must now be aligned to the molding machine. This is accomplished primarily through use of the locating ring and sprue bushing. While these components are both part of the mold, they interact with the molding machine to create the necessary alignment between mold and machine. The purpose of the locating ring is to position the mold on the machine platen accurately in preparation for attaching holding clamps, as shown in Figure 7-11.

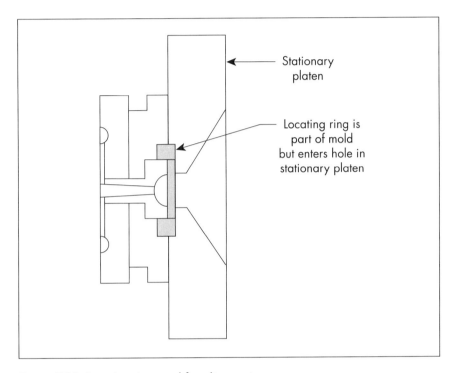

Figure 7-11. Locating ring used for alignment.

Locating rings are included in the purchase of a complete mold base from a mold base supply company, but are also available as individual components as replacement items. The outside diameter ranges from approximately 4 to 5.5 in. (101.6 to 139.7 mm), with the most common diameter being 4 in. (101.6 mm). The inside diameter matches the head diameter of the sprue bushing being used for a specific mold, which is normally 2 in. (50.8 mm) in diameter. The ring depth is recessed into the top plate of the mold base by .218 in. (5.5 mm) and rises from the top plate surface by the same dimension. However, there are some design concepts that require the stand-up dimension to be different from .218 in. (5.5 mm).

The outside diameter of the locating ring forms a slip fit with a matching hole located in the stationary machine platen to which the mold is mounted. After the mold is located using this locating ring device, it must be "squared" to be parallel to the earth by using a liquid level placed across the top plate. The mold is rotated slightly around the locating ring until the leveling monitor indicates parallelism.

The purpose of the sprue bushing is to position the nozzle of the molding machine (which is adjustable) accurately to ensure proper fit and sealing of the nozzle to the mold, as shown in Figure 7-12. This allows molten plastic to enter the mold under the sustained pressures created by the molding machine. If the sprue bushing is not correctly aligned, and of the proper size, material leakage will occur and result in a loss of pressure used for filling the mold. In addition, this leakage will travel back along the nozzle and cover the nozzle heater band, which usually results in a damaged nozzle heater.

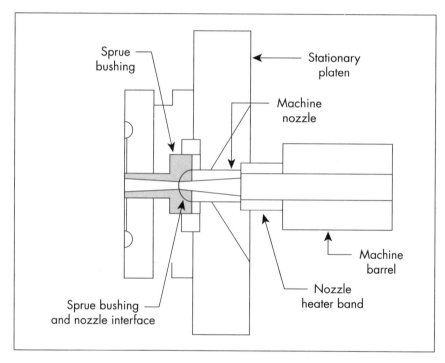

Figure 7-12. Sprue bushing used to position nozzle.

Sprue bushing dimensioning is covered in another section of this book, so we will not repeat that information here. However, we can state that sprue-bushing dimensions are critical for achieving proper molding process conditions.

Platens, Tie Bars, and Platen Bushings

The machine *platens* are large plates of steel used for mounting and supporting the injection mold and for opening and closing the mold during production. There are actually three platens in the standard machine. Two of these are stationary and tied together with large diameter bars, known as *tie bars*, as shown in Figure 7-13.

In between the two stationary platens is a moving platen. This platen travels along the tie bars by way of close fitting bushings that are located in each corner of the moving platen. The injection mold is located between the stationary platen found at the injection end of the machine and the moving platen. It is clamped in place there, and is opened and closed through the action of the moving platen, positioned by way of a hydraulic or toggle mechanism at the clamp end of the machine. The structural integrity of the platens and tie bars provides a sturdy and accurate alignment mechanism to ensure the mold is always properly located and positioned in reference to the machine during the entire injection molding process.

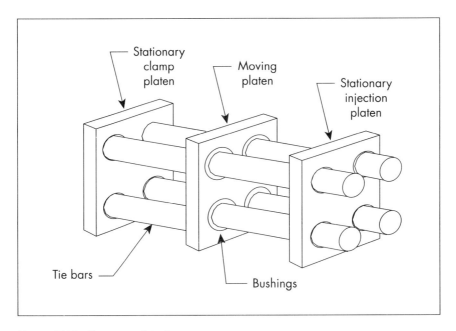

Figure 7-13. Platens and tie bars.

The degree of parallelism between all platens determines the machine's ability to properly locate and close the mold. This parallelism is controlled by way of micrometer thread adjustment on the nuts that hold the platens to the tie bars. The faces of the platens must be ground flat and parallel to very tight tolerances, and kept in good repair to maintain their parallelism. In addition, the tie bars must be checked frequently to ensure they are straight and parallel, and the tie bar bushings must be replaced as soon as they are worn. The bushings are lubricated to minimize wear.

To properly monitor and adjust for parallelism and alignment throughout the entire machine and mold combination, the frame of the machine must be leveled so it is parallel to the earth. This is accomplished through the use of vibration pads placed on the floor of the molding room and under specific leveling points of the molding machine. The vibration pads are adjustable in height to accommodate the leveling process and absorb much of the vibration generated by the molding machine during the molding process.

SUMMARY

To help provide high-quality products at the most efficient cost, the injection molding process requires precise alignment of the mold components to each other and of the mold to the molding machine.

Mold halves are commonly aligned using leader pins on one half that mate with bushings on the other half.

Tapered locks assist in aligning mold halves for products that require extreme accuracy of molded dimensions.

Slides, cams, and lifters are action devices that require locks and alignment methods to ensure location, as well as prevent damage to the mold.

Telescoping components are available for ensuring alignment of critical mold elements.

Alignment of the mold to the molding machine is accomplished through use of a locating ring and sprue bushing.

Platens require sophisticated machining to achieve the parallelism required for precise machine alignment to the mold.

QUESTIONS

1. What is the definition of alignment as it applies to this chapter?
2. What is the common way to align the A and B plates of a mold?
3. Why is one leader pin unit offset in the mold?
4. Why should leader pins be placed as close to the edge of the mold as possible?
5. What is the common term that applies to moving components within a mold's interior?
6. Name the three most common action devices.
7. What are the primary components involved with aligning the mold to the molding machine?
8. How many platens does the standard horizontal molding machine have?
9. What determines the accuracy capability of the molding machine to properly locate and close the mold every cycle?
10. Besides absorbing machine vibration, what other part do vibration pads play concerning the machine frame?

Repairing, Protecting, and Storing Molds

METHODS OF REPAIR

It is safe to state that any given mold will require a certain amount of repair for damage occurring during the use or storage of that mold. The more it is used, or the longer it is stored, the more damage will occur. Damage that occurs during use includes items such as broken core or ejector pins, peened parting lines, and worn gate areas, while damage most common during storage is in the form of rust.

It is a good idea to save the last shot from any production run and keep it with the mold in storage. This provides a visual example to the mold maintenance area of how the parts were being produced. A repair person can inspect the parts to determine the fitness of the parting line, cavity surface condition, ejector pin position, and other pertinent information. A written statement of problems seen by the molding room personnel should also accompany this last shot.

What Causes Damage?

Improper care is a major cause of damage to molds. The use of metal screwdrivers to remove stuck parts results in scratching the cavity surfaces. This scratching causes appearance defects on the molded part, but may also affect a specific dimension and cause it to be out of tolerance. It may also act as an undercut and cause subsequent parts to hang up and not eject, or at least crack. Failure to lubricate moving components, such as slides and cams, or leader pin bushings, will result in a galling of the sliding components that will eventually cause the components to seize. Even wiping the cavity surface with an improper rag can cause slight to major damage to the highly polished surface. Improper processing can cause major damage, such as when a process technician uses too much injection pressure at start up and flashes the mold. That flash can force its way between leader pins and bushings, or down ejector pinholes, or into vertical separations, and can force the sidewalls out or move around slide mechanisms and cams. This can lock the mold in place as it tries to open. Allowing the nozzle to drool can result in material oozing onto the A side of the cavity set. This is usually thin material that solidifies quickly and becomes a hardened sheet of plastic. If it is allowed to stay, it can cause heavy damage to the cavity set when the mold closes on it. If mold temperatures are not properly controlled, a portion of the mold can run at a higher temperature. If the difference is greater than 10° F (5.5° C), there is potential for thermal expansion variations that cause swelling and galling.

After a mold has completed a specific number of production cycles (varying with type of product and material being molded), it must be cleaned to remove residue formed during the molding process. The cavity surfaces and vent areas will show the greatest amount of residue, but the ejector housing, ejector pins, and runner blocks are other areas that also collect residue. Because of the tight tolerances being maintained by the mold, this residue buildup may become enough to keep the mold from closing properly and/or force specific dimensions out of tolerance. Vent areas especially require attention. If the vents plug (due to residue), they become ineffective and higher injection pressures will be required to fill the mold. This may cause flashing, which may result in mold damage.

When the mold is pulled from the molding machine, it should be thoroughly cleaned, inspected, and coated (primarily inside, but lightly outside) with a rust preventive material to minimize the possibility of damaging rust formation. The coating should be especially heavy for long-term (over 30 days) storage. It is important to clean out the waterlines and coat them. If possible, molds should be stored in a cool, dry area to minimize rust-producing conditions.

Repairing damage to a mold can range from being a simple process to one of tremendous difficulty, depending on the degree and type of damage. In some cases, a repair may require only the replacement of a broken ejector pin, while in other cases, the entire cavity set may need to be replaced. The method of repair, then, depends on what caused the damage. Removing rust from the outside surface of a mold is simple and usually requires no more than a wire brushing for even extreme cases. However, removing rust from a cavity surface is a totally different situation and may even require welding and refinishing damaged areas. Alternatively, the rust may have caused so much damage that the cavity sets and other components may have to be replaced completely.

Molds are expensive, custom-made products and can be damaged by many things. They require a great deal of care in their operation and storage phases. Preventive maintenance, in the form of cleaning and lubrication, (and coupled with proper care and processing) will go a long way toward minimizing damage.

Designing for Repair and Maintenance

It is advised that, for high volumes or long runs, the mold be designed so that vulnerable components (such as gate areas) can be easily replaced or repaired. This can be accomplished by the use of inserts and laminated construction. While this may be more expensive to incorporate at the initial building stage, it provides a much simpler way of repair when such is needed, and downtime will be minimized. In the end, the overall costs of mold maintenance are much lower. If a mold is designed for ease of repair, it will last much longer because welding will not be needed for repairs. Welding can greatly reduce the life of a mold because of stresses created during the welding process.

In building molds that contain "actions," it is advisable to incorporate readily accessible lubrication points (such as grease fittings) into the action design to make it easy for the molder to lubricate them on a routine basis. High volume molds should also incorporate lubrication grooves in action areas as well as leader pins and guided ejector plates to capture grease for extended lubrication efficiency.

Repairs by Welding

Beyond simple repairs performed by replacement of damaged ejector pins and other components, welding is the most common method of repairing mold damage. This is due to the relatively short period of time required for welding a repair versus other methods. However, welding can be extremely detrimental to the overall life of a mold because of the thermal stresses that are induced on the mold steel.

Welding should be considered a combination of scientific techniques and personal art. One cannot be used for mold damage repair without the other. Poor workmanship (the "art" side) is responsible for most weld failures, and can cause cracking or warping of the mold or its components that will seriously jeopardize the basic function of the mold during the molding process. Welding is often used to correct small defects, such as localized cracks or dents or chipped areas. Sometimes it is used to modify an area because of an engineering change requirement. Although welding is the most common method of repair, it really should be used only as a last resort. However, after it is determined that welding will be utilized, it must be done with utmost caution and care.

There are two basic types of welding commonly used in the mold repair business: basic welding with coated wire electrodes and tungsten inert gas (TIG) welding. The coated wire process can be used for large area welding, such as repairing cracked mold bases or plates. However, it is not suitable for correcting small cracks or parting line areas. Use of the TIG process in those areas results in a finer weld structure and less possibility of metal stress, because the TIG process results in lower heats and faster cooling times.

Regardless of the welding method utilized, basic rules-of-thumb apply.

- The welding rod material should be of the same composition as the mold material being welded. This will minimize stress caused by reactions to subsequent heat treating processes.
- To increase hardness and create a fine (rather than coarse) weld structure, welding amperage should be kept to the low side.
- The entire mold component being welded should be brought to recommended preheat temperature (see metal suppliers information) and kept at that temperature during the entire welding process.
- The preheat temperature must be set to be above the martensite-forming temperature of the metal being welded. This information is available from the metal supplier.

- To maximize weld strength, all tool steels must be cooled to at least 212° F (100° C) after welding, and then normalized or annealed according to the metal supplier's recommendations.

Welding of aluminum is also possible, but requires a few modifications to these rules. Please contact your supplier for correct welding procedures.

Metal Deposition Process

Another common method of repair to small areas of minimal thickness is the metal deposition process (commonly called the Selectron process after the company that commercialized it). This process allows damage repair to areas that are up to approximately .15 in.^{2} (1 cm^2) in area and from .04 μin. to .04 in. (1 μm to 1 mm) in depth. The process is similar to standard welding and entails connecting a cathode to the component to be welded. The anode consists of a holder with a cotton swab attached. The cotton swab is dipped in a suitable liquid electrolyte and brushed across the treated area until the desired thickness is attained.

The metal deposition process also can be utilized to apply various metallic coatings, such as gold, nickel, copper, or cadmium to mold surfaces. These would be applied to provide corrosion protection, as well as to correct dimensions of components such as bushings and core pins. In most cases, subsequent machining is not required.

PROTECTING THE MOLD

Molds are expensive tools. They require many hundreds (and possibly thousands) of hours of machining and crafting before they are ready to be used for production purposes. Damaged molds result in losses to the molder and customer, not only in repair costs, but in lost production and machine downtime. Although molds are usually made of high-grade steel, they are easily damaged, especially the cavity image areas and parting lines. For this reason, molds should be treated with the utmost care and protected during the production process, as well as during storage between runs.

New Molds

New molds demand just as much protection as older molds. The first item on the protection agenda for a new mold should be to open it at the parting line. Pry bar slots should be machined into two opposing corners on one half of the mold at the parting line to allow insertion of pry bars for ease of opening. Figure 8-1 shows the location of these slots.

After the mold has been opened, an inspection should be made of all exposed surfaces to ensure no obvious damage has occurred during shipment. Cavity image surfaces should be explored in fine detail, looking for peened parting lines, scratched surfaces, chipped core pins, missing ejectors, etc. After this inspection is completed, the sprue bushing and locating ring should be examined to make sure they fit the intended machine and are of the proper size for the plastic that

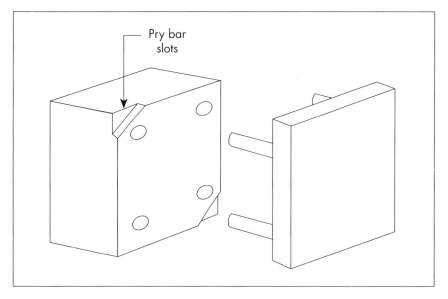

Figure 8-1. Pry bar slots.

will be molded. Sprue bushings must have a smooth and polished internal tapered opening. Any blemishes will cause problems when molding.

Actions, such as slides, cams, lifters, and ejector systems, should be activated by hand to make sure they are operating smoothly and properly. Cutting oils and lubricants must be cleaned away and an inspection should be made for metal chips, filings, and dust. Dirty cutting oils and lubricants will cause scratches and/ or galling of metal surfaces and must be removed. Make sure vents are present and cleaned out.

The entire mold, including external surfaces, waterlines, actions, vent areas, and cavity surfaces must be carefully cleaned and sprayed with a protective finish. The specific protective finish used will depend on the length of time that will pass before the mold is sampled or run. Short-term storage (up to 3 days) only requires a light spray of vegetable oil or mild rust preventive (including petroleum jelly). But longer-term storage (7 to 30 days) will require a heavier coating, and very long term storage (30 days or longer), a special coating that is heavily applied. The outside surfaces also should have rust preventive applied. Waterlines should be sprayed with rust preventive for short-term storage, and filled with readily soluble foam for long-term storage.

Molds in Production

While molds are being run in production, they are exposed to many situations where the potential for damage is high. To begin with, when a mold is placed and located in the molding machine, it may come apart prematurely and fall from the machine if connecting straps are not used to keep the two mold halves together during handling. In addition, during this initial installation, items such as water-line connections, special electrical connectors, slide actuators, and other external components are subject to damage from accidental impact with other items, such as the press frame and platens. The locating ring can be nicked or peened if not correctly aligned and may deform enough to not properly fit the mating platen hole of the machine. Burrs and nicks should be removed and polished out.

When the clamping mechanism is brought forward for adjusting to the newly installed mold, it should be done slowly and allowed to gently squeeze the mold tight. Overclamping is a common cause of mold damage, especially to fragile items such as small ejector pins and cores.

After the mold is ready for clamps, they should be installed so that they intentionally "toe in." This means the clamp should have its heel adjusted to point the toe slightly (.125 in. [3 mm] is fine) toward the platen as shown in Figure 8-2. This must be done because it is impossible to maintain exact clamp parallelism to the platen for maximum clamping force. Expansion and contraction of the mold and machine result in clamps slipping loose when adjusted to be perfectly parallel. If the clamps are adjusted so the toe is pointing away from the platen, the clamping force is also pointing away from the platen, and the mold may fall out due to insufficient clamp force. Therefore, the toe should be adjusted to point in toward the platen to ensure that the clamping forces are directed toward the platen. Then, the clamps should be tightened to the torque recommended for the specific bolt size being used.

Installing and Setting Up the Mold

This section will discuss the sizing and inspection, and installation procedure for the mold. The steps used for properly mounting a mold and starting it up for production are critical. Following these steps will ensure proper mold installation and protection.

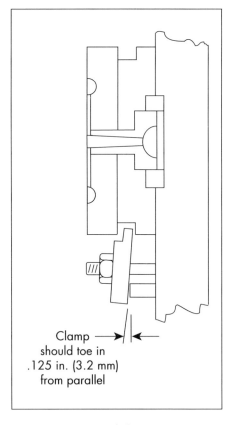

Clamp should toe in .125 in. (3.2 mm) from parallel

Figure 8-2. Toe in of clamps.

Sizing and Inspection

A machine must be selected that is properly sized for the specific mold being installed. After the machine is selected, it must be inspected to determine its status. This inspection includes items such as:

- Proper hydraulic oil level.
- Heater bands in place and operating.
- Mold temperature controllers operable.
- Empty injection cylinder with screw forward.
- Hopper shut-off closed and hopper wiped clean.
- Proper material available and dried.
- Granulator clean and available.
- Safety gates and mechanisms operating and in good condition.
- Vent hoods clean and operating.
- Heat exchanger clean and operating.
- Machine lubricated or auto-lubrication working and filled.
- Alarms and lights operable.

Installation Procedure

After the machine inspection is completed, the mold can be installed. The following steps should be taken, but they are generic in nature and do not preclude following the machine manufacturer's instructions. Always follow those instructions first.

1. Make sure that the mold has a connecting strap installed. This strap should connect the two halves of the mold and keep them from coming apart during transportation. Normally, this is a metal strap mounted across the A and B plate parting line. It is not a safe or proper practice to install the mold as two separated halves.

2. Start the machine, make sure the injection sled is in the full back position, and set the barrel heaters to the proper temperatures. The profile should run from a cool rear setting to a progressively hotter front zone and nozzle as outlined on the setup sheet. Turn on the feed throat cooling water.

3. Open the clamp wide enough to accept the mold. Normally, this dimension equals a minimum of twice the height of the mold. This may require resetting the mold open limit switches or control settings. Refer to the machine manual for instructions.

4. Lower the mold from the top of the machine (or slide it in from the side) using a chain fall, and bring the mold up against the stationary platen by hand. The mold should rest against the platen without assistance. This is accomplished by adjusting the location of the chain fall toward the platen. It is a good idea to have a thick metal plate placed across the lower tie bars at this point. The plate will act as a safety catch in case the chain fall breaks or the connecting hook opens.

5. At this point, the mold must be raised and lowered slightly in an attempt to position it so the locating ring on the mold will slip into the locating hole on the platen. The chain fall should be connected so that the mold tilts slightly at the top.

6. The tilted mold should be placed slightly above the locating hole of the stationary platen, and held against that platen as the mold is slowly lowered. The locating ring of the mold will automatically slip into the locating hole of the platen as the mold is gently lowered. A level can be placed across the top of the mold to assist in aligning the mold so it fits squarely on the platen. After leveling, the A half of the mold is ready to clamp in place.

7. Position clamps, adjust, and bolt the A half of the mold to the stationary platen. The mold should be mounted with at least one clamp in each of the four corners. If the mold is very wide, additional clamps should be placed along the long dimension. If the mold is very small, it may be possible to use only two clamps per mold half, although this is not recommended; smaller, specially built clamps may have to be utilized to use four clamps per mold half. The best thing to do is purposely adjust the heel of the clamp away from the platen. This results in the clamping forces being directed toward the platen through the toe of the clamp. The angle of direction should be minimal and can be set so that the clamp heel is only .125 to .250 in. (3.2 to 6.4 mm) away from parallel. The clamp is slotted for linear adjustment to allow the clamping bolt to be located as close as possible to the mold, because that, too, aids in creating maximum clamping pressure on the mold itself.

8. If ejector rods are required, place them in the mold now. Then, slowly bring the clamp unit forward, under low pressure, to prepare for clamping the B half of the mold. This may require adjusting limit switches or settings. Check the machine manual for this information. Bring the moving platen up to within .25 to .50 in. (6.4 to 13 mm) of the mold base and set limits for the "high pressure close" to activate at that point. Now, continue moving the clamp unit forward until it touches the mold base. Allow the press to build up clamp pressure to the desired setting. This ensures that the mold is fully closed.

9. Shut off the machine. Locate clamps, adjust, and bolt the B half of the mold to the moving platen, making sure to follow the procedure mentioned earlier.

10. Remove the chain fall hook, eyebolt, and connecting strap from the mold. To avoid losing the connecting strap, it may be desirable to keep it mounted, but swung out of the way and tightened down so it will not come loose and cause mold damage.

11. Recheck each clamp on both halves to make sure they are all tight. Start the machine and *slowly* jog the clamp unit open under low pressure,

watching for any indication of the mold halves seizing or binding together. Open the mold approximately .5 in. (13 mm) and stop. Shut off the machine and fully check the mold to make sure it is properly mounted.

12. Start the machine and continue to open the mold slowly until the B half disengages fully from the A half. Then stop the mold at the point described for fully open. This would normally be a minimum of approximately two times the depth of the part being molded, to make sure the part will fall free after ejection. It is acceptable to open the mold farther than that rule-of-thumb, but it should not open farther than necessary because of the additional time required to do so. If there are slides or other actions in the mold, make sure they are still properly engaged upon full opening, if at all possible. This will minimize the potential for breakage. Check for broken springs or other obvious damage.

13. Adjust settings for proper ejection. Ejection should *not* pulsate. One stroke should be adequate for part removal. If this is not enough, there is something wrong and it should be corrected before continuing production. The amount of ejection stroke should not exceed 2.5 times the depth of the part in the B half of the mold (assuming ejection is located on the B half). The ejection stroke should be just enough to get the plastic part in the B half freely out of the mold. More than that only adds to the overall cycle time.

14. Lubricate all moving components, such as ejector guides, leader pins, and slides. Wipe off all excess lubricant. Gently clean the cavity surfaces. Close the mold and turn off the machine.

15. Attach the hose lines from the mold temperature control unit. Blow air through the cooling lines of the mold to make sure they are not obstructed and to observe the proper path for connecting hoses. Do not loop the A half and B half together on a single line, but attach separate in and out lines for each half, and use separate control units for each half. Make sure there are no kinks in the hoses and that they will not be crushed or stretched when the mold closes or opens. After inspecting for proper attachment, activate the temperature control units and adjust for the proper temperature setting.

16. Recheck all clamps.

17. Check to determine if the barrel is up to heat. It normally takes 45 to 60 minutes to properly come to preset temperatures and soak. Make sure all heater bands are operable and properly connected.

18. Ensure that the hopper feed gate is closed and the hopper magnet is in position. Place fresh, dry material in the hopper. A purging compound may be required first, depending on what material was in the barrel last.

19. After the barrel is up to heat and allowed to soak for 10 to 15 minutes, open the feed gate on the hopper and allow material to drop through the feed throat and into the cylinder.

20. Purge the machine as follows:
    a. The injection screw should have been left in the forward position of the barrel when the last job was shut down. It should stay in that position while preparing material for air shots. Activate the screw rotation until fresh material is brought to the front of the barrel. This will be obvious because the screw will spin freely to begin with, but slow considerably as the fresh material is brought forward.
    b. Set the screw return limits to the desired point and allow the screw to return to that point. The screw rotation will stop after the screw returns to the set point. Allow the material that was brought forward enough time to absorb heat from the cylinder. This will normally be only a minute or two.
    c. With the sled still in the back position, take three air shots. An air shot consists of injecting a full shot of material into the air, under molding pressure, and allowing it to accumulate on a special plate designed to catch purgings. Make sure that proper time is allowed between these air shots to allow the upcoming material to come to proper heat. This time usually amounts to the total cycle time of the job that will be running in production. Using a fast-acting pyrometer with a probe, measure the melt temperature of the material injected during the air shots. This temperature must be controlled for proper molding. Adjust settings as necessary. If a different material or color is being used, it may require 15 or 20 air shots to clear the old material out, or a purging compound might be required.
21. Set all limits for injection and cycle. These include injection forward speed and pressure, holding pressure, cushion distance, cooling time, mold open and close settings, and others as required.
22. Prepare a full charge of material for the first shot. Bring the injection sled forward until the nozzle seats against the sprue bushing of the closed mold. The mold must be closed to absorb the force of the injection sled against the sprue bushing. If the mold is open at this point, the A half may be pushed off the platen. Lock the sled control in place.
23. Open the mold and bring the clamp unit to a full open position.
24. Set the cycle indicator to manual, semi-automatic, or automatic, depending on requirements.
25. Close the safety gate to initiate the first cycle.
26. Observe the injection process. The pressures and feeds should be set so that a short shot will be taken first. Then pressure and feed settings can be adjusted until a properly molded part is produced. This should be done over a long period of time (15 to 20 shots) and not hurried.
27. During production, a mold must be monitored (even if it is running on automatic cycle) to make sure ejected parts do not stick in the mold, and

that all actions are operating properly. Periodic cleaning and lubrication should be performed as often as two or three times a shift, depending on the type of material being molded and the complexity of the mold design. Cleaning should be performed gently, with a cleaner designed specifically for molds. If agitation is required, such as in vent areas, a light colored nylon-scrubbing pad should be utilized. The light color designates a mild abrasive action as compared to an aggressive abrasive action found in the darker colors. Scouring powders and steel scouring pads should never be used because they will damage the surface of the mold.

Mold releases should be discouraged, but if they are absolutely necessary, use them sparingly. It is common to think that if a little is good, more is better, but that is definitely not the case concerning mold release. If mold releases must be used, an investigation should be held to determine why, and the cause should be corrected as soon as possible. Mold release acts to "gum up" the works, especially the slides, cams, and lifters. This can result in a great deal of damage to the mold.

If the mold must run for a long period (such as 30 days or so) it should be stopped periodically and cleaned, lubricated, and inspected. Waterlines should be disconnected and tested to make sure they are clean and operating properly. Parting lines should be inspected for peening. The sprue bushing should be checked for nicks and burrs, as should the entire mold.

After the production run is completed, the mold should be shut down. This includes saving at least the last shot to come off the mold. This is done for inspection purposes later, so the mold repair personnel can determine what, if anything, must be done to prepare the mold for the next run. This last shot should include the entire tree, including the runner and sprue, with parts still attached, if possible.

Next, the mold should be thoroughly cleaned and sprayed with rust preventive, including the waterlines. It is a good idea to send the mold to the tool room for inspection, final cleaning, and final rust spray before placing it in storage.

STORING MOLDS

Between production runs, molds are usually taken from the molding machine and placed in some type of storage. The storage area normally consists of a series of heavy-duty metal racks onto which the molds are placed and tagged for identification. Usually, the mold has specific identification stamped on at least two adjacent sides so it is readily visible, and the rack has an identification tag mounted on each station that corresponds to the mold identification. This helps keep everything in order and in preparation for subsequent use.

The type of mold protection required during storage depends on the length of time needed. Long-term storage requires a heavy-duty procedure that results in a heavy coat of protective material being placed on the mold. This coating is difficult to remove. A much lighter coating can be applied for short-term storage.

Short-term Storage

Short-term storage can be considered storage that will be for only a day or so, all the way up to 30 days. For this type of storage, the mold must be cleaned thoroughly and then sprayed with a neutralizer to remove any acids present from fingerprints or other acid-producing items. Then, a coating of rust preventive should be applied, a light coat for a few days storage and a medium coat for up to 30 days storage. The waterlines should be blown dry with compressed air and then sprayed with a rust preventive. The outside of the mold should be wiped clean and sprayed lightly with rust preventive or light oil. The inside of the sprue bushing should be sprayed with light oil also.

It may be necessary to tear down the mold and clean up all the ejector pins, return pins, cavity surfaces, runner blocks, etc., especially if corrosive plastics were being molded. The mold components should be inspected for damage, cleaned, and lubricated for storage.

Long-term Storage

Long-term storage can be defined as storage lasting over 30 days in time. If it is unsure how long a mold will be in storage, assume it will be for longer than 30 days. This way the mold will be sure to be protected, regardless of the time span.

For long-term storage, the mold definitely should be torn down and inspected. Damage should be repaired and the mold components and base should be thoroughly cleaned, neutralized, reassembled, and then coated inside and out with a heavy coating of rust preventive. It should be assumed that all the steps mentioned in short-term storage also apply here. However, the waterlines should be filled with protective foam that is available from mold base supply houses. This prevents the mold from rusting from the inside out. The foam rust preventive material is of much heavier duty and longer lasting than general rust preventive material.

Some companies insist that all long-term molds be sealed in plastic wrap before being placed on the storage rack. However, this may result in trapped condensation. Normally, it is better to have dry air circulating around the molds to help prevent rust.

SUMMARY

It is safe to state that any given mold will require a certain amount of repair for damage occurring during its use or storage. The more it is used, or the longer it is stored, the more damage will occur.

Improper care is the major cause of damage to a mold.

Mold damage repair can range from being simple to that of great difficulty, depending on the degree and type of damage.

For repair and maintenance purposes, it is advised that molds be designed so that vulnerable components (such as gate areas) can be easily replaced or repaired. This can be accomplished by the use of inserts and laminated construction.

For other than simple repairs, such as replacement of damaged ejector pins, welding is the most common method of repairing mold damage. However, due to the stresses that are created, welding should be used only as a last resort.

Molds are expensive tools and should be treated with the utmost care and protected during the production process, as well as during storage between runs.

Even the process of installing a mold in the molding machine for production can result in damage if it is not performed properly.

Mold releases should be discouraged because they tend to cover up a potential mold-damaging situation. They also cause slides, cams, and lifters to become sluggish and may result in costly damage to those components.

Short-term storage requires the use of a light-duty rust preventive, but long-term storage (over 30 days) requires the use of a heavy-duty rust preventive not only on the mold, but also in the waterlines.

QUESTIONS

1. Why should the last shot be saved from each run?
2. What can be considered the major cause of damage to a mold?
3. What can be done at the mold design stage to allow ease of repair later?
4. What can be incorporated in actions to minimize damage?
5. How can welding be detrimental to the overall life of a mold?
6. What are the two types of welding commonly used for mold repair?
7. How can the hardness of a weld be improved?
8. What is considered to be long-term storage?
9. Why should a connecting strap be used on a mold?
10. Should mold clamps be adjusted toe in or toe out, and why?
11. Where should the screw be left when a job is shut down?
12. Describe an air shot.

Troubleshooting Product Defects Caused by Molds 9

OVERVIEW

In many cases, the mold designer and moldmaker are not responsible for troubleshooting defective plastic parts. However, this chapter may still be useful because it explains the primary cause of some common mold-related defects and their solutions. The first two books in this series, Volume I: *Manufacturing Process Fundamentals* and Volume II: *Material Selection and Product Design Fundamentals,* both have troubleshooting chapters with a different focus. For more information, it may be useful to consult them.

Too often, when technicians, engineers, or operators are presented with a part (or set of parts) with defects deriving from molding problems, they start turning dials, flipping switches, and adjusting timers, without understanding what they are doing or knowing what results to expect. This is a common occurrence and has its genesis in the practice of troubleshooting a defective part by way of doing something (anything) that worked in the past when a quick fix was desired (but not possible). It does not have to be that way. The situation should be such that the troubleshooter can objectively analyze a molding defect and eventually determine a possible solution. The solution should be attempted, followed by another decision. If the first solution does not work, another solution should be devised and attempted. However, each solution should be determined independently and rationally. There should be no guesswork, and assistance from outside sources should be welcomed and pursued if necessary.

One common source of troubleshooting assistance comes from material suppliers. They usually have a convenient laminated guide sheet available that tells what to do if certain defects are encountered. These guide sheets are well researched and a troubleshooter may eventually find the answer to a specific problem, but may never know what caused the problem or why the particular solution actually worked. The guides do not go into that kind of detail. If they did, they would be written like a book. SME (Society of Manufacturing Engineers) has a three-tape video set (entitled *Troubleshooting Injection Molding Problems*) that graphically demonstrates the causes and solutions to the most common defects. It is also available as an interactive CD-ROM.

The best approach to troubleshooting is to use as many of these sources as possible, and mix in a good dose of common sense.

WHAT CAUSES DEFECTS?

A study that took place over a 30-year span (1963 to 1993) by Texas Plastic Technologies analyzed the root causes of most common injection molding defects. Because troubleshooting takes effect after acceptable parts are molded, the study only investigated defects that were process-related and did not include those resulting from poor design of the product. It was determined that the defects could be traced to problems with one or more of the following items: the molding machine, the mold, the plastic material, or the molding machine operator. Most interesting was the percentage that each of these items contributed toward the cause of the defects. Figure 9-1 summarizes the results.

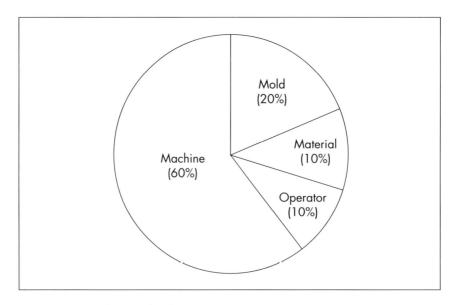

Figure 9-1. Distribution of defect causes.

Most of us in the industry believe that the most frequent cause of defects is the material, with the operator coming in a close second. However, as Figure 9-1 shows, the most frequent cause of defects is the molding machine. Thus, in troubleshooting, the first place to look for a solution to a defect problem is the machine, because the answer will be there six out of 10 times. If the solution cannot be found there, the next place to look is the mold.

A troubleshooter must be able to approach a problem with an objective mind. What solved a problem one day may not solve the same problem another day. Because of the large number of parameters, the variables of these parameters, and the way they all interact, many solutions may exist for a single problem. Likewise, many problems may be fixed using a single solution. So, the troubleshooter must think through the problem and make sure the proper solution is chosen by applying objectivity, simple analysis, and common sense.

The first key is to visualize the way a process should be running. Most troubleshooting is actually performed after a specific job has been running successfully for an extended period. There has been an initial setup and debugging process and the mold has been accepted for production. Then, after running successfully, parts begin to be molded with defects. This is when the troubleshooter is introduced into the picture. This is also when common sense and objectivity must be brought into play.

COMMON DEFECTS AND REMEDIES

On the following pages are listed 24 of the most common molding defects. There is also a listing of the most common mold-related cause of each defect along with the most popular remedy suggestions. The list of defects is not intended to be all-inclusive, nor is the list of causes and remedies. This section is only intended to assist the mold designer or moldmaker in understanding some of the potential mold-related defects that may show up in parts produced on their molds.

Black Specks or Streaks

Sprue Bushing Nicked, Rough, or not Seating

If the sprue brushing is nicked, rough, or not seating, it may cause the material to degrade. The reason for this is that the material is trapped in residence (until it overheats) in an area of the mold that retains high heat. The degraded resin becomes brittle and will break loose, entering the melt stream and showing up as black specks or streaks.

Inspect the internal surface of the sprue bushing. Remove any nicks or other imperfections. Check with thin paper or bluing ink to see if the radius of the nozzle is equal to (or smaller than) that of the sprue bushing.

Burned Material Caused by Improper Venting

Improperly vented areas show up with a whitish ashing on the mold steel or a charring of the plastic material in that area. This is from the burning of air trapped in the mold and compressed to the point of ignition.

Locate vents in the mold at points 1 in. (25.4 mm) apart all along the perimeter of the cavity image on the B half of the mold. If more venting is possible, do it. There is no such thing as too much venting.

Blisters

Inadequate Temperature Control

The mold must be run within a specific temperature range for a specific material. Inadequate cooling line distribution will result in some areas of the mold cooling faster than others. These cooler areas will cause the plastic to skin over in those locations while the rest of the plastic is still solidifying. Air and gases will form pockets in these areas and result in a blister.

Make sure that waterlines are located as close to the cavity image as possible and that they closely follow the cavity image contour. The idea is to create even distribution of heat throughout the mold, but especially in the cavity image area.

Insufficient Venting

Proper venting is a very important part of building a mold. The correct size, location, shape, and number of vents need to be considered and analyzed in the mold design stages. If insufficient venting exists, trapped air cannot escape from the mold and will form pockets that evolve into blisters.

Vents should be placed at 1 in. (25 mm) intervals along the perimeter of the parting line. In addition, the runner should be vented the same way. Air that is trapped in the runner will be pushed into the cavity and collected there.

Blush

Sprue Diameter too Small

If the sprue diameter is too small, it causes material to *freeze off* (solidify) too soon after it enters the mold and can result in blush, especially in a sprue-gated part. Blush located at a surface gate indicates the gate is probably too thin, resulting in the same effect.

Make sure the sprue diameter at the nozzle end is large enough. This information can be obtained from the supplier of the plastic being molded. Stiff materials (like polycarbonate) require large diameters for easier flow. Also, check the gate depth. The material supplier will provide the range of depth required for a specific material. Be sure a rectangular gate has generous radii in the sharp corner areas.

Bowing

Improper Gate Location

Often, the cause of bowing in a molded part is that the gate has been located in a thin section feeding a thick section. This results in uneven molecular sizing due to the tendency for the material to cool down and solidify as it tries to travel through the thinner section first. When shrinking, the molecules in the thicker section shrink more than the molecules in the thinner section, and this causes the part to twist and bow.

Gates should be located with two thoughts in mind: gate into the thicker sections, and centralize the flow of material. Gating into the thicker section allows the material to completely fill the mold before it cools and shrinks. A centralized flow allows an even distribution of material through the cavity image. The closer these rules-of-thumb are followed, the less bowing will be present.

Inconsistent (or Uneven) Mold Temperature

A molded part must be allowed to cool in a mold having consistent temperatures. If there are "hot spots" in the mold, those areas will be the last to cool and the resultant differential in shrinkage between those hot spots and other areas will result in distortion of the part (*bowing*).

Design and construct the mold so that both the A and B halves have the same capacity for cooling. This will minimize the tendency for the plastic part to stay on whichever half is hotter. In addition, design the cooling channels to ensure that there will be no more than a 10° F (5.5° C) difference between any two spots of the mold.

Brittleness

Gate and/or Runner Restrictions

Gates and runners that are too small, or have excessive sharp corners, may cause a shearing and tearing of the plastic as it flows through them. This will result in a separation of layers in the laminar plastic flow. These separated layers will cool quickly and not be able to knit back together. The result will be weak molecular bonding exhibited as brittleness in the part.

Examine the gates and runners to ensure they are built to the recommendations of the material supplier. Every plastic family has a specific requirement concerning gate and runner design, and these should be followed as closely as possible.

Bubbles (Voids)

Section Thickness too Great

When a plastic part consists of varied wall thicknesses (instead of one steady thickness), the thicker walls will cool (and solidify) last. There will be a pressure loss in those thick areas as they continue to cool after the thinner areas have solidified. The plastic will pull away toward the solid section and cause a void in the thick section. When the void is on the surface of a part, it appears as a sink mark. When it is below the surface, it appears as a bubble.

The best solution (although expensive) is to use metal core-outs to thin the thicker wall. Alternatively, if possible, change the wall thickness so that the thicker section is no more than 25% thicker than the thin section. This will minimize the void.

Burn Marks

Improper Venting

Venting systems are placed in molds to exhaust any gases or trapped air that might be present. If the vents are not deep enough, or wide enough, or if there are not enough vents, the air is compressed before it is all exhausted and it ignites and burns, charring the surrounding material.

Vents must be a minimum of .125-in. (3-mm) wide. The vent land should not be more than .125-in. (3-mm) long. Blind areas, such as the bottom of holes, should have vents machined on the side of ejector pins that are placed there. There should be enough vents on the parting line to equal 30% of the distance for the parting line perimeter. Thus, a 10-in. (25-mm) long parting line perimeter would have 12 vents, each of which measures .25-in. (3-mm) wide (3 in. [76 mm] total). Another rule-of-thumb states that a vent should be placed at 1-in. (25-mm) intervals around the parting line perimeter.

Clear Spots

Cracked Mold

A possible source of clear spots in a molded part is a cracked mold or cavity set. If the crack emanates from a waterline, moisture can be seeping into the cavity and it can be trapped as water droplets, appearing as clear spots in the molding. The seepage may not be visible with the mold open and may only occur when the mold is closed and under clamp pressure.

Cracked mold bases can be repaired, but the cause should be determined. It is possible that a waterline was placed too near a cavity set, thereby weakening the steel between them. Alternatively, a cavity set may have been weakened during the hardening process. Of course, there are a variety of possible causes. It may be necessary to replace the entire mold. If a waterline has cracked open, it is possible to insert a copper tube through the waterline and use it as the cooling channel as a temporary repair. If welding is used to repair cracked cavity sets or mold bases, the welding should be performed only by a reputable welder, experienced in this specialized type of welding.

Cloudy Appearance

Uneven Packing of Cavity

Uneven packing can normally be traced to improper gating, runner sizing, or location. The material enters the cavity at the wrong spot, which does not allow the material to be packed against the mold steel in all areas. The material solidifies without replicating the mold finish, and this appears as a cloudy area. There is also a possibility that one area of the molding surface was not polished as well as the other areas. This would give a cloudy appearance in that area.

Make sure that the mold is properly polished. If so, investigate the proper gate size, number, and location for a specific product design and material from the material supplier.

Contamination

Excessive Lubrication

Molds containing actions such as slides, cams, and lifters, require periodic lubrication to ensure continuous production. However, in some cases, the molder finds it difficult to get to lubrication points and overloads them when they are lubricated. Excessive lubrication can move into the cavity image area and contaminate the molded part.

Design and construct actions with lubrication points and fittings that are easily accessible to the molder. This will encourage the molder to lubricate only when required and reduce the amount of lubricant used.

Cracking

Insufficient Draft or Polish

Draft angles should be an absolute minimum of 1° per side to facilitate easy removal of the part from the mold. Ejector pressure may cause cracked parts if less than that is used. In addition, rough cavity surfaces (and other undercuts) cause a drag on the part as it ejects. This may cause cracking if the ejection pressure is increased to push the part over this rough surface.

All sidewalls should be drafted to the highest degree possible, but to 1° minimum. Check for, and remove, undercuts formed by peened parting lines or other shut-off areas. Cavity surfaces should receive a high polish when the mold is built and re-polished as the need arises.

Crazing

Because crazing is simply a very fine network of cracks, the same causes and remedies apply that are mentioned under "cracking."

Delamination

Excessive Mold Release

If a mold release is required at all, it is necessary to limit its use. Too much mold release will cause the mold release to penetrate the molded layers. This will keep the layers from bonding and result in delamination. Keep mold release away from presses unless necessary and then only use as a temporary fix until the cause of the sticking can be rectified.

Check the mold and sample parts to determine why a part is sticking. Look for, and repair, undercuts formed by damaged parting lines. Check core pins, looking for mushroomed ends. Determine if polishing is required to help the molded part release from the mold.

Discoloration

Improper Mold Temperature

In general, a hot mold will cause the material to stay molten longer and allow the molecules to pack tighter. The result is a very dense part that appears darker due to that density. On the other hand, a cold mold will cause a loss in gloss because the material cools before it can be forced against the mold surface. This will cause a less dense part, which will appear lighter.

The waterlines should be large enough to accommodate the correct flow of water needed to provide turbulence with a Reynolds value of between 3,500 and 5,000. A lower value will result in improper water flow and inconsistent temperature control. If the flow becomes laminar, the mold temperature will fluctuate during the run and result in parts of varying shades of color.

Flash

Nonparallel Shut-off Lands

Over time, the primary plane of the parting line may develop an out-of-parallel condition. This could be caused by many reasons, including cavity sets that have coined into the mold base, or flash that has been allowed to sit on the parting line plane, eventually forming depressed edges.

Examine the parting line, looking for damaged areas, especially those caused by loose flash or drooling of the nozzle. These must be repaired or replaced. Also, look for sunken cavity sets. Cavity sets must sit above the mold surface by .002 to .005 in. (0.05 to 0.127 mm) to effectively stop flashing of the parting line.

Inadequate Mold Support

Molds must be properly supported behind the B plate to compensate for the open space caused by the U-shaped ejector housing. Without proper support, the injection pressure pushing molten plastic into the mold will cause the B plate to flex, resulting in flashing at the parting line.

Use at least one row of support pillars to provide support to the backside of the B plate. See Chapter 3 for information on support pillars and support plates.

Flow Lines

Inadequate Venting

Flow lines may be the result of trapped air and gases keeping flow fronts from packing together. The laminar style of flow that exists in the molten plastic material requires separated layers of materials to knit together to form a structurally sound product. Trapped gases reduce the capability of achieving this throughout the part, resulting in flow lines.

Vent the mold to ensure a vent at 1 in. (25 mm) intervals along the parting line perimeter. Vent the runner too. If in doubt, vent it. Vents can be added to ejector pins to release trapped air from deep pockets or bosses. Vents should even be added in the gate area. The material supplier will provide proper venting information for specific materials.

Gloss (Low)

In general, the same conditions that cause "cloudy appearance" and "blush" defects may contribute to low gloss appearance. The remedies are the same.

Jetting

Excessive Gate Land Length

The small area that encompasses a gate is called its *land*. It determines the total distance the molten material must travel in a restricted state before it is allowed to enter the cavity itself. If the land is too long, the plastic material will start to cool too quickly and will have to be forced into the cavity. This action causes the flow front to split apart and allow fresh material to flow between the splits, causing the classic snakelike appearance on the surface of the molded part.

Decrease the land length so it is no greater than .125 in. (3.18 mm) but no less than .031 in. (0.79 mm).

Knit Lines (Weld Lines)

Difficult Product Design

Knit lines are the result of a flow front of material being injected at an obstruction in the mold cavity. This obstruction is usually a core pin being used to create a hole in the molded part. The flow front breaks up into two separate fronts and goes around the obstruction. When the two fronts meet on the other side, they try to weld back together again (knit) and form a single front once more. Another cause of knit lines is the use of more than a single gate, which results in multiple flow fronts. As the number of gates are increased, more knit lines are added.

After the conditions exist that create a knit line, it can not be eliminated. It can only be manipulated. Moving the gate moves the knit line. Increasing melt and mold temperatures help to minimize the knit line. By strategically adding gates, knit lines can be made to reside in specific locations. For example, a core pin can be shortened to allow flash to form over its face. Then, the knit lines can be caused to form in the flash. The flash (and the knit lines trapped in the flash) can then be removed from the part and discarded.

Nonfill (Short Shots)

Insufficient Venting

Venting is used to remove trapped air from the closed mold so molten material will be able to flow into every section of the mold. If the air is not removed, it acts as a barrier to the incoming plastic and will not allow it to fill all sections of the mold. The result is nonfill. The mold should be vented even before the first shot is made. Vent the runner first, and then create enough vents on the parting line to equal 30% of the length of the perimeter surrounding the cavity image. An additional approach is to use a vacuum system in the mold to help pull the trapped air out before injecting material.

Shrinkage (Excessive)

Improper Dimensional Calculations

There are over 20,000 plastic materials to choose from for molding a product. Each has a specific shrinkage factor assigned to it. The moldmaker must incorporate that factor in each dimension of the part so when the part is ejected from the mold it will shrink to the correct dimensions. It is necessary for the mold designer and moldmaker to understand the particular shrinkage characteristics of the material being molded. This information is available from the material supplier and should be closely heeded. Even then, if a material is changed or a different grade is used, it may not shrink the way that was predicted.

It is wise to leave all dimensions "steel safe" if possible. This will allow modifications to be made if the shrinkage does not come out the way it was predicted. Most molds are not even expected to produce perfectly acceptable parts on their initial trial. Usually, final dimensions are negotiated between what the product designer intended and what the mold actually produces. Critical dimensions must be developed in the mold by the moldmaker until they produce the desired requirement.

Sink Marks

Unbalanced Flow Patterns

Aside from the obvious causes of sink marks (such as thick walls where ribs meet connecting walls or inefficient cooling), one major cause of sink marks on parts produced in multiple cavity molds is an unbalanced runner and gate system. The material may enter each cavity at a different time and cause differences in the exact moment at which the material begins to solidify. This may result in shrinkage differences that may cause sink marks to appear.

The flow path should be constructed so that every cavity *finishes* filling at the exact same moment. This is the true definition of a *balanced runner system*. When this is achieved, all the material in the entire shot will stop moving at the same time and begin to solidify and shrink at the same rate. This will help minimize excessive shrinkage conditions that, in turn, will minimize sink marks.

Splay (Silver Streaking)

Gates are too Small

When material enters the cavity, it must do so in a laminar fashion (layer upon layer). If the gate is too small, the flow front going through may be broken up into many small fronts, and the material will enter the cavity in a spray pattern instead of the desired laminar pattern. This spray pattern is duplicated on the surface of the molded part in the appearance of splay.

Gates should be examined to ensure they are of the correct size and shape for the specific material being molded. The material supplier is a good source for that information. During the examination, look for obvious problems, such as parting line burrs or peened gate edges. These too will cause the flow front to break up into small spray patterns.

Warpage

Nonuniform Ejection

The ejection system of the mold may be worn (or improperly supported) causing uneven ejection of the molded part. If the part is not evenly ejected, warpage may result as the still warm part is distorted while it is being pushed out of the mold.

Check the operation of the ejection system. Molds that are planned for long-term use should incorporate a guided ejection system, which consists of guide pins and bushings. This practice makes the mold's ejection system much more durable and long lasting. Without this guidance, gravity forces cause the ejector system to wear out the standard supporting holes and the plates will sag, causing uneven ejection.

Bibliography

Bryce, D. Plastic Injection Molding series. Volume 1: *Manufacturing Process Fundamentals.* Volume 2: *Material Selection and Design Fundamentals.* Dearborn, MI: Society of Manufacturing Engineers, 1996, 1997.

Crackwell, P.S. and Dyson, R.W. *Handbook of Thermoplastics Injection Mould Design.* New York: Blackie Academic and Professional, 1993.

Fundamentals of Tool Design, Fourth Edition. Dearborn, MI: Society of Manufacturing Engineers, 1998.

Groover, M. and Zimmers, E., Jr. *CAD/CAM:Computer-Aided Design and Manufacturing.* Englewood Cliffs, NJ: Prentice-Hall, Inc., 1984.

Injection Molding Technology. Seventeen videotapes. Chester, CT: Paulson Training Programs, Inc., 1993.

Jacobs, P.F. *Stereolithography and other RP&M Technologies.* Dearborn, MI: Society of Manufacturing Engineers, 1995.

Mitchell, P., ed. *Tool and Manufacturing Engineers Handbook.* Volume 8: *Plastic Part Manufacturing.* Dearborn, MI: Society of Manufacturing Engineers, 1996.

Morena, J.J. *Advanced Composite Mold Making.* New York: Van Nostrand Reinhold, 1988.

Oberg, E., et al. *Machinery's Handbook,* 25th Edition. New York: Industrial Press, 1996.

Troubleshooting Injection Molding Problems. Three-tape video set or interactive CD ROM. Dearborn, MI: Society of Manufacturing Engineers, 1996.

Appendix A

An Injection Mold Design Check List

The following items are not listed in any particular order, nor do they constitute an all-encompassing list. They are to be used only as a guideline to ensure that important details are at least considered prior to finalizing the mold design.

1. Make sure mold layout places cavities as close as possible to the center of the mold (sprue) to minimize flow travel distance.
2. Specify maximum draft allowable.
3. Use dowels to align cavities and cores.
4. Make sure cooling channels (waterlines) are close enough to cavity contour for proper cooling, but not too close to cause breakthrough. Ideally, cooling channels should be located to be twice their diameter from the cavity.
5. Ensure use of support pillars where practical.
6. When possible, support core pins in opposite half of mold to minimize deflection during injection.
7. Specify a shut-off land around cavity inserts to minimize clamp pressure requirements.
8. Place cores and cavities in pockets to minimize shifting.
9. Design cavities and core blocks as sections (rather than cut in the solid) to simplify machining, add strength to the total assembly, and provide venting paths for trapped air.
10. Check the mold for the press in which it is intended to run, to ensure proper die height capacity, ejector system location and operating method, capacity of injection unit, capacity of clamp unit, and ease of installing mold.
11. Place cores in the ejector half if possible to facilitate ejection of molded part.
12. Provide holes for eyebolts (or other lifting mechanisms) for ease of installation and removal from press.
13. Provide lifting straps to keep mold halves together during transport.
14. Double check shrinkage factors for plastics being molded.
15. Maintain uniform wall thickness by including cored areas.
16. Use guided ejector systems for long-run projects.

17. Utilize lubrication fittings for any actions and make them easily accessible.
18. Identify waterlines such that the "in" is at a lower level than the "out."
19. Indicate a vent for every in. (2.5 cm) of parting line perimeter.
20. Indicate those areas where gate and ejector pin marks cannot be allowed.
21. Identify any textured area requirements and include the texture identification.
22. Make sure pry bar slots are incorporated in opposing corners of the mold at the parting line.
23. Specify all materials to be used in the mold construction and their respective hardness and finish requirements.
24. Perform calculations to determine proper waterline diameter based on Reynolds number values.
25. Calculate gate size based on formulas or finite element analyses.
26. Calculate runner sizes based on location and distance of resin travel.
27. Determine number of cavities based on annual volume requirements and number of months desired for running annual production needs.
28. Attempt to place ejector pins at high-strength locations such as beneath bosses or in corners of the molded part.
29. Design high-wear areas, such as tunnel gates, as insertable components to facilitate ease of repair.
30. Place identification numbers in each cavity of multiple-cavity molds.
31. Obtain written customer approval for high-concern items such as gate locations, ejector pin markings, cavity numbering, and texture.
32. Initiate discussions with the selected moldmaker to ensure understanding and agreement on mold design issues.
33. Check for unnecessary, and undesirable, vertical shut-off areas.
34. Make sure sprue bushing and locating ring are properly sized for resin requirements, as well as for the molding machine being used to run the mold.
35. Have the final design checked and approved by the customer, moldmaker, and others deemed necessary prior to the start of building the mold.

Appendix B

History of Plastics

HOW IT ALL BEGAN

In 1868, an enterprising young gentleman by the name of John Wesley Hyatt developed a plastic material called celluloid and entered it in a contest created by a billiard ball manufacturer, that was held to find a substitute for ivory which was becoming expensive and difficult to obtain. Celluloid was actually invented in 1851 by Alexander Parkes, but Hyatt perfected it to the point of being able to process it into a finished form. He used it to replace the ivory billiard ball and won the contest's grand prize of $10,000, a large sum in those days. Unfortunately, after the prize was won, some billiard balls exploded on impact during a demonstration (due to the instability and high flammability of Celluloid) and further perfection was required in order to use it in commercial ventures. But, the plastics industry was born, and it would start to flourish when John Wesley Hyatt and his brother Isaiah patented the first injection molding machine (1872), in which they were able to injection-mold celluloid plastic. Over the next 40 to 50 years others began to investigate this new process for manufacturing such items as collar stays, buttons, and hair combs. By 1920, the injection molding industry was off and running, and it has been booming ever since.

Charting Industry Evolution

The following time table shows some of the important dates regarding the evolution of the injection molding industry:

Table of Evolution

1868—John Wesley Hyatt injection molds Celluloid billiard balls.
1872—John and Isaiah Hyatt patent the injection molding machine.
1937—Society of Plastics Industry founded.
1938—Dow invents polystyrene (still one of the most popular materials).
1940—World War II events create large demand for plastic products.
1941—Society of Plastics Engineers founded.
1942—DME introduces stock mold base components.
1946—James Hendry builds first screw injection molding machine.
1955—General Electric begins marketing polycarbonate.

1959—DuPont introduces acetal homopolymer.

1969—Plastics land on the moon.

1972—The first parts removal robot is installed on a molding machine.

1979—Plastic production surpasses steel production.

1980—Apple uses ABS in the Apple IIE computer.

1982—The JARVIK-7 plastic heart keeps Barney Clark alive.

1985—Japanese firm introduces all-electric molding machine.

1988—Recycling of plastic comes to age.

1990—Aluminum molds make their mark in high-volume production injection molding.

1994—Cincinnati-Milacron sells first all-electric machine in the United States.

1996—Two-platen machines are introduced in the United States.

Answers to Chapter Questions

CHAPTER 1

1. The cavity image.
2. Many functions and features can be incorporated into the design of the product.
3. Less than a minute.
4. Between the A and the B plates.
5. The sprue bushing is the component that allows molten plastic to enter the mold and begin its travel to the cavity image.
6. This is accomplished through use of a runner system and gates.
7. It is theoretically possible to fill any size or shape cavity image with only a single gate.
8. This is an unsafe practice because the side rails of the box will distort under injection and clamp pressure and the box may crack.
9. The primary difference is that an ejector pin contacts the surface of the plastic part or runner, while the return pin contacts only a steel surface of the mold.

CHAPTER 2

1. Crystalline materials.
2. Low shrink is .000 in./in. to .005 in./in., medium shrink is .006 in./in. to .010 in./in., and high shrink is anything over .010 in./in.
3. The melt index (or melt flow or flow index), ASTM # D1238.
4. Because cavity number will determine the size of the mold and the size of the equipment needed to run the mold.
5. 1040 carbon steel.
6. A-1 for the smoothest, and D-3 for a coarse, dry-blasted finish.
7. Because it may create a vacuum between the plastic and the metal surface of the mold.
8. The part will hang on to the ejector pin and may not fall out of the mold when ejected.
9. The ideal layout would be like spokes in a wheel.
10. The ideal runner cross section is round because this results in equal pressures in all directions.
11. By utilizing vents (venting).
12. Theoretically, a mold can last forever.

CHAPTER 3

1. The application of injection and clamp pressures.
2. The combination of core and cavity.
3. Free standing.
4. Pocketed cavity sets.
5. For securing the cavity set in the pocket.
6. Two times the waterline's actual diameter.
7. Up to 10 tons or more of pressure.
8. 14 in.^{2}
9. 56 in.^{2}
10. The ejector housing is a one-piece, U-shaped component.
11. Electrical discharge machining.
12. Stereolithography apparatus.

CHAPTER 4

1. An action.
2. A horn pin.
3. Because it may hit on the slide or cam surface when the mold closes, causing damage.
4. Slides and cams.
5. Lifters.
6. 25°.
7. Body, head, and face.
8. Due to tolerances required, dimensional stackups, and expansion/contraction of the mold plates during the molding process, it is impossible to maintain an original flush condition for ejector pins.
9. Too short a pin.
10. A stripper plate (also, a sleeve ejector).

CHAPTER 5

1. Runner, gate, vent.
2. Flash will form between the two and the sprue will stick in the sprue bushing.
3. It captures the first (and coldest) portion of the incoming plastic and keeps it from plugging up the flow path in the runner system or gate areas.
4. Because it creates equal pressure in all directions.
5. A diameter of .125 in.
6. 20%.
7. Because it utilizes the straight line runner approach and minimizes the travel any of the plastic flow fronts must make to get to the cavity images.
8. Cycle times are shorter because there is no runner thickness to consider, defects are minimized as a result of minimizing stresses, and scrap is reduced because there is no runner to dispose of.

9. In the thickest wall section going from thick to thin.
10. The melt flow index, ASTM # D1238.

CHAPTER 6

1. Waterlines machined throughout the mold to allow water to flow through the mold.
2. 10° F (5.5° C).
3. Heated water from the first half will then be used to try to cool the second half.
4. There is no way of knowing without checking the mold with a pyrometer.
5. With turbulent flow the water is constantly being tumbled and mixed which results in all of the water being in contact with the mold metal at one time or another.
6. 3,500.
7. No more than 10° F (5.5° C).
8. Waterlines should be located as close as possible to the surface of the mold actually forming the molded part.
9. Copper alloys have five to six times the thermal conductivity of common mold steels.

CHAPTER 7

1. Alignment can be defined as the accurate locating of components to their respective positions.
2. Through use of leader pins and bushings.
3. To ensure that the mold halves can only be assembled to each other in a specific manner.
4. To allow maximum area for placement of cavity images and waterlines.
5. They are called actions.
6. Slides, cams, and lifters.
7. The sprue bushing and locating ring.
8. Three. Two are stationary and tied together with tie bars, and the third is movable and travels along these tie bars.
9. The degree of parallelism between the moving and stationary platens.
10. They also are used for leveling the machine.

CHAPTER 8

1. The last shot provides a visual sample of how the part was running.
2. Improper care.
3. The mold can be designed such that all vulnerable components (such as gate areas) are made as inserts which can easily be replaced.
4. Lubrication points that are easy to access.
5. By inducing thermal stresses in the mold steel.

6. Basic welding with coated wire electrodes and TIG (tungsten-inert gas) welding.
7. Keep welding amperage to the low side.
8. Storage of longer than 30 day's time.
9. To keep the two mold halves together during handling.
10. Toe-in to direct the clamp holding force towards the platen.
11. In the forward position.
12. An air shot consists of injecting a full shot of material into open air, with the injection sled back, onto a special purging plate, under normal injection pressure.

Index

N

O

P

Q

R